The European Right to Be Forgotten

The European Right to Be Forgotten

The First Amendment Enemy

Kristie Byrum

LEXINGTON BOOKS
Lanham • Boulder • New York • London

Published by Lexington Books
An imprint of The Rowman & Littlefield Publishing Group, Inc.
4501 Forbes Boulevard, Suite 200, Lanham, Maryland 20706
www.rowman.com

Unit A, Whitacre Mews, 26-34 Stannary Street, London SE11 4AB

British Library Cataloguing in Publication Information Available

Library of Congress Cataloging-in-Publication Data

Names: Byrum, Kristie, author.
Title: The European right to be forgotten : the first amendment enemy / Kristie Byrum.
Description: Lanham, Md. : Lexington Books, 2018. | Includes bibliographical references and index.
Identifiers: LCCN 2018022769 (print) | LCCN 2018023050 (ebook) | ISBN 9781498549615 (electronic) | ISBN 9781498549608 (cloth : alk. paper)
Subjects: LCSH: Right to be forgotten—European Union countries. | Privacy, Right of—European Union countries. | Electronic data interchange—Law and legislation—European Union countries. | Data protection—Law and legislation—European Union countries. | Records—Law and legislation—European Union countries.
Classification: LCC KJE1626 (ebook) | LCC KJE1626 .B97 2018 (print) | DDC 342.2408/58—dc23
LC record available at https://lccn.loc.gov/2018022769

™ The paper used in this publication meets the minimum requirements of American National Standard for Information Sciences—Permanence of Paper for Printed Library Materials, ANSI/NISO Z39.48-1992.

Printed in the United States of America

For my family,

Reed, Bradley, Caroline, James, Donald,
Peggy, Charlotte, and Lily

Contents

Acknowledgments

I wish to acknowledge my husband, Reed Bolton Byrum, a diligent editor and constant supporter during the authorship of this book. My research team, Donald J. Baumgartner and Elizabeth Ann Sabol, dutifully navigated structure, editing, and deadlines throughout the process. Librarians at Bloomsburg University of Pennsylvania offered valuable research suggestions and assistance. Artist Marisa Peters created an illustration and veteran book author Katharine Nohr provided useful advice and friendship during the writing process. Lastly, I thank scholars Victor J. Vitanza and Kathleen Donohue Rennie for encouraging advanced academic endeavors.

Preface

This text examines the implementation of the European Right to Be Forgotten, a part of the European General Data Protection Regulation that sanctions removal of information from search engines, allowing for the creation of "memory holes." The book addresses the specter of a similar adoption in the United States and explains how the right to be forgotten would controvert the legality of the U.S. First Amendment and other constitutional protections. The book reviews multiple historic and contemporary dimensions that demonstrate the inherent conflict of the European right to be forgotten with the free flow of information in the United States. Though privacy activists in Europe and the United States may position the right to be forgotten as a protection of individual rights, First Amendment advocates point to the U.S. Constitution's Bill of Rights as a shield against such government intervention, censure, and regulation in online content. While the current debate encompasses conversations about digital privacy, data protection, and individual rights, the U.S. dialogue requires distinctly different parameters. The First Amendment remains as a beacon guiding discussions, enlightened by the heritage of the free flow of information and the expected future of transparent communication. The First Amendment stands up against government interference, corporate actors acting as data controllers, and encumbrances on principles of transparency. As a former newspaper journalist and public relations executive in the United States, I present this text from a perspective grounded in journalism and mass communication education, earned in the United States, strengthened by decades of news-gathering and content creation, and bolstered by First Amendment advocacy for the robust free flow of information in society. The point of the view contained in this text rejects the right to be forgotten as effrontery to the tenets of the First Amendment for protection of freedom of speech and freedom of the press. This book proposes evidence of the conflict between the right to be forgotten with the free flow of information in the United States and offers proof points found in U.S. Supreme Court decisions, constitutional theories of the First Amendment, and contemporary standards of practice. Reflecting a media scholar's perspective, this book provides clarity into the imperative for the free flow of information in society, not only in the United States but also across the globe, and gives international audiences a thoughtful critique, as they grapple with complex issues associated with the right to be forgotten. Today, the availability of accurate information flow suffers relentless siege from "fake news," citizen journalism, foreign government hacking, absence of accountability, and technology-driven campaigns to falsify legitimate information. The use of computer bots and the ability for foreign governments to place advertisements on Facebook and other media channels to influence democratic elections has now come to light in the United States, suggesting new global forces in elections by virtue of technology and new media channels that were not available before. The right to be forgotten represents just one of many forces influencing the availability of online information, and the First Amendment's freedom of speech and freedom of press provisions face threats from a multitude of directions. Falsification of news, decline of objective media, and deterioration of media as watchdogs on government present chilling contemporary First Amendment threats. Further, ambivalence by the audience poses a troubling circumstance, because uncertain citizens fail to question the authenticity and correctness of news and information offered to the masses. While media's continuous adaptation is a relevant concern, these topics require in-depth analysis in other venues. This book explains why the right to be forgotten conflicts with the First Amendment and addresses contiguous concerns about online privacy, storage of data on the web, and the conflation of government intervention and regulation by private enterprise. The advent of new technologies and emergence of government regulation and intervention in routine information dataflow will continue to dominate global agendas because of the enduring gravitas of information that influences social, economic, and political concerns worldwide.

Introduction

Just view these startling contemporary headlines: The *Washington Post* reports Russian meddling in United States 2016 Presidential Election (Shapiro, 2017). A leading trust study indicates that social media platforms and search engines contribute to the fact that the media is the least trusted institution globally (Edelman, 2018). The U.S. Justice Department indicts 13 Russian citizens who sought to subvert the election to favor Donald Trump by creating false social media accounts and purchasing paid advertisements to discredit opponent Hillary Clinton (Apuzzo & LaFraniere, 2018). These headlines mark a significant deterioration in accurate information flow in the U.S. democracy at the hands of foreign actors who sought to meddle with political speech. What happened in the election is not prior restraint or censorship; instead, Russian agents presented false statements masqueraded as social media posts from fictitious individuals and organizations to persuade American voters. These circumstances create an alarming backdrop for the United States, and freedom of information undergoes threats from many directions, including the European right to be forgotten. We must now thoughtfully consider the vital role search engines play in information delivery in democracies.

This book addresses how implementation of the European General Data Protection Regulation's right to be forgotten encumbers the free flow of information in U.S. society, and, while the right is allowed in Europe, the notion encounters staunch opposition in the United States. The right to be forgotten, which allows for the removal of data from search engines in Europe, collides with United States constitutional principles and unearths a myriad of global media law issues. The notion of the right to be forgotten places privacy supporters and freedom of information advocates at sharp odds, calling into question the digital practice of information freedom in democratic societies. The role of technology and corresponding search engines emerges into keen focus, when we consider the World Wide Web as a vast repository of information used to garner information for decision-making by democratic societies and their citizens. In this Digital Age, the ability to access the web, utilize reputable search engines, and engage with the world around us is integrated into living, learning, and global citizenship. A study of global Internet users claimed 98 percent of Internet users agreed or strongly agreed that the Internet is essential for their access to knowledge and education (Internet Society, 2012). Consistent with this thinking, 83 percent of the same respondents agreed Internet access should be considered a basic human right (Internet Society, 2012).

This book demonstrates why the right to be forgotten raises multiple concerns in the United States and how U.S. law stands in opposition to universal data restrictions on information retrieval across all Google domains. The arbitrary creation of a private corporate "data controller" with the power to de-link and thus "censor" information insults the First Amendment by precluding the free flow of information and endangers the robust debate serving as a foundation for democratic principles historically found in the United States. On a global basis, many media and communications scholars endorse the notions of free flow of information in society, authentic and accurate disclosure, and adherence to principles of transparency. Today journalists, media scholars, legal analysts, and private corporations struggle to understand how accurate transparent communication could occur in an ecosystem that allows arbitrary information removal. This book indicates the European notion of the right to be forgotten challenges U.S. constitutional law by imperiling the marketplace of ideas and eroding disclosure of information to the public. This work addresses the reasons why the right to be forgotten will falter in the United States by revealing not only the contravening constitutional elements of the First Amendment, but also by showing how the right to be forgotten violates First Amendment theories and how it collides with core values that advocate for transparency, disclosure, and free flow of information.

A leading exponent of the seventeenth century's Age of Enlightenment, philosopher John Locke wrote about man's susceptibility to law from three sources: the divine, civil, and the societal law of "opinion" or reputation (Locke, 2001). Perhaps Locke's "law" of preserving the reputation functions as the driver for the right to be forgotten and foretells the European response to the right to be forgotten as a privacy concern. Yet in the Digital Age, the intimate details of reputations swarm through wireless networks, rapidly projected across the globe. The European right to be forgotten allows for dissembling of materials about an individual in the country where the de-linking is requested, yet data may still appear in the U.S. search engines, diminishing the impact of the de-linking in a global economy. This text explores the implications of the European right to be forgotten from legal, ethical, and social perspectives. These chapter summaries provide a preview of the content and arguments contained in this book.

CHAPTER 1: THE LANDSCAPE

To analyze the impact of the right to be forgotten on the free flow of information—and why the U.S. Constitution's First Amendment precludes the removal of information through censorship—we must first understand the history of the Bill of Rights and the introduction of the First Amendment as a protection against government intervention. The words, "Congress shall make no law respecting an establishment of religion, or prohibiting the free exercise thereof; or abridging the freedom of speech, or of the press; or of the right of the people peaceably to assemble, and to petition the government for a redress of grievances," (Legal Information Institute, 2018) has withstood the enduring test of time undergoing numerous disruptions in technology, social pressures, and changing political circumstances. The First Amendment, grounded in a solid foundation to prevent government intervention in the affairs of religion and freedom of expression, provides citizens with a stalwart protection against government-sanctioned censorship and restraints on media. The advent of the European right to be forgotten stems from an alternative source of law, preoccupied with the privacy of individuals in a digital world, and one that tacitly allows infringements on freedom of information. Thus, the notions of freedom of expression and privacy become convoluted in the Information Age. A central point of difference between Europe and the United States stems from diverging views and cultural values attached to privacy. In the United States, the right to inform often supersedes the individual's privacy right in a free and open media environment. Article 8 of the European Convention of Human Rights clearly specifies that, "everyone has the right to respect for his private and family life, his home and his correspondence" (European Convention of Human Rights, 2017). Despite the clear solemnity of this right, the administration of this right may conflict with other rights, most notably freedom of expression. European Courts have attempted to resolve conflicts over these rights, often favoring individual privacy rights over freedom of societal expression, particularly when private individuals litigate to preserve privacy (Dulong de Rosnay & Guadamuz, 2016). In contrast, the United States legal system typically adheres to the cultural values of an uninhibited "right to inform" and places the First Amendment principles above the privacy protection when disputes find their way to the courtroom. To date, the bulk of literature on the right to be forgotten focuses on understanding the legal rationale for the right to be forgotten (Ambrose, 2016;

Bunn, 2015; Lindsay, 2014), the corresponding notions of oblivion, digital redemption and forgetting (Ambrose, Friess & Van Matre, 2012; Blanchette & Johnson, 2002; Gorzeman & Korenjof, 2017; Hoskins, 2014), censorship (Dulong de Rosnay & Guadamuz, 2016) and the encumbrances on freedom of information (Kulk & Borgesius, 2014; Larson, 2013).

The implementation as the European General Data Protection Regulation of 2018 across the European Union has created a circumstance whereby an individual can request that a search engine de-link information about her or him with the contention that it is outdated and no longer relevant, dispatching that information into "oblivion" and confining it to a state of "forgetting." In what has been called "digital redemption," privacy activists applaud the European right to be forgotten, while others dismiss the notion as an infringement on the free flow of information of society. For some, digital redemption is viewed as a joyful and just action, while for others it represents an assault on the accuracy of history, impairment of the marketplace of ideas, and a generator of web-based memory holes—symbols of author George Orwell's dystopian world of the future in the book *1984* (Orwell, 1949). In Europe, information may be removed by filling out an online form that includes identity information. Since the onset of the right to be forgotten in 2014, this activity has caused the elimination of information about hundreds of thousands of individuals. News websites such as the BBC and *The Daily Telegraph* have experienced de-linking at the hands of Google, Bing, and other search engines because of European government regulation that forced elimination of hyperlinks that met the standards for removal. While the right to be forgotten thrives in Europe, the threat to the free flow of information has now begun to cross global borders. This book demonstrates how the right to be forgotten imperils information flow, historical clarity, and accuracy thus tilting toward censorship. The book offers a counterpoint that reveals how the First Amendment provides clear constitutional protections against the implementation of the European right to be forgotten in the United States.

CHAPTER 2: THE MARKETPLACE OF IDEAS: THE DISRUPTION OF DYNAMICS

U.S. jurisprudence has adopted the concept of the "marketplace of ideas," transmitting concepts first articulated by John Milton in 1644 in *Areopagitica* into the contemporary landscape. Milton wrote this tract to protest against censorship and today it has become a well-regarded essay for the defense of freedom of expression. The World Wide Web's marketplace of ideas operates 24 hours a day, seven days a week. While Europeans value tenets of privacy over the free flow of information in society, American audiences operating in the environment where the First Amendment exists appreciate the freedoms molded into the First Amendment of the Bill of Rights adopted by the country's founding fathers. In distillation, the First Amendment prohibits the passage of any laws that would encumber freedom of speech and freedom of the press. Therefore, under the most rigid interpretation of the law, the First Amendment would preclude the implementation of the right to be forgotten in the United States. By examining theories of the U.S. First Amendment in tandem with the rationale for the European right to be forgotten legislation, a conflict of principles and practice rapidly emerges. This discussion may gain enlightenment from the theoretical framework of the marketplace of ideas that has been central to the U.S. Supreme Court arguments for preservation of freedom of speech and press since the nation's inception. The marketplace of ideas theory presumes ideas should possess the opportunity to compete in the marketplace in an effort to gain acceptance. By applying the marketplace of ideas concept to the web's search engine, a myriad of questions emerge. We can now contemplate whether the marketplace of ideas operates properly, offers accurate information, and provides access of the market's information to humans operating with rational thought. This text asserts the right to be forgotten disrupts the information marketplace by compromising the inventory of information available online. In the United States, corresponding movements to de-link information from common search engines like Google taint the First Amendment and arbitrarily consolidate power of the information in a central information source, effectuated in many cases by a private individual making a la carte information decisions, based on variable rules. The implementation of such a system warps truth, and the available information for citizens to glean truth in this circumstance is, in fact, dictated by a private power source. In the case of the right to be forgotten, the European Union has enacted a law that transfers regulation of the activities to private companies, in this case the search engines exercise discretion over the information available on a given subject. This government intervention led by the European Union leads to incomplete search engine results, delivered with multiple holes. The cascading results of government intervention and information regulation, for example, propel the systematic handicapping of journalists who may attempt to investigate a story or present information into the marketplace for consideration by citizens. Ordinary information-seekers suffer too, because the marketplace of ideas is compromised and once-valued transparency now gives way to a murky sea of information that has some tidbits of retrievable data but is lacking possible salient information simply gone forever, digitally erased after search engine de-linking. This book calls for the preservation of the information marketplace and implores government regulators and private corporations to keep the marketplace of ideas open for online visitors, consequently providing citizens, institutions, and journalists availability to relevant information.

CHAPTER 3: COMMUNICATION IN DEMOCRACY

Chapter 3 addresses the role of communication in democracy by evaluating the function of the free flow of information in educating citizens. The availability of social, economic, and political information in a representative democracy enables the formulation of an informed electorate. Further, with the benefit of information the electorate is thus positioned to discern truth, debate ideas, and arrive at a public opinion by virtue of the free exchange of information in the society. The chapter addresses the tenets of Alexander Meiklejohn, who espoused a compelling need to protect the free flow of information in society—particularly in the realm of political speech—because of its fundamental contribution to self-governance. In a Meiklejohnian sense, political speech ascends to a preferred position because of its significant role in communicating information to voters and catalyzing the formulation of public opinion. Scholars have attempted to bring democratic elements forward to the Digital Age, as evidenced by scholarly work that considers contemporary democracy amid the cacophony of information provided on search engines.

In a further exploration of media practice, the chapter articulates the traditional watchdog role of the media that keeps a providential eye on the political sphere and paves the way for investigative journalism that exposes a variety of ills influencing the public in a democracy. The traditional watchdog function presses for accountability in the social, economic, and political spheres. This important news media role, historically admired and respected by citizens, suffers under the right to be forgotten. The action of search engine de-linking impairs the ability of the news media to deliver relevant information to the citizenry, so the electorate may perform its democratic role to form an opinion and exercise the right to vote. The right to be forgotten can alter accurate reflections on an individual, a societal occurrence, or a trend in the culture. Facts about specific individuals and happenings in the culture become warped through the distortions of the right to be forgotten, which leaves holes in the collective understanding of events. In addition to the First Amendment protection of free speech, United States law provides other legal protections, including the Freedom of Information Act, providing journalists and the public access to federal records, and the Government in Sunshine Act, ensuring the access of reporters to public meetings where elected officials discuss and construct laws and policies that influence the lives of the citizens. These legal measures indicate the social value awarded the free flow of information in society and the assurances for transparency designed to enable education of citizens in the democracy. Citizens have embraced a common belief that they have "a right to know," and, for the online sphere, they have come to interpret that right as the ability to gather, understand, and store information about individuals. The Digital Age creates a new topography for the free flow of information in a democratic society. Today, traditional journalists working for print, broadcast, and online news media encounter an evolved dynamic—the citizen communicator publishing information directly to other consumers. Elected officials also communicate directly with their key publics, disrupting the news media delivery channel involving gatekeepers who make decisions about the scope and treatment of the news. These circumstances provide the backdrop for the contemplation of the right to be forgotten and enlighten further exploration to verify why any movement to incorporate the right to be forgotten in the United States conflicts with the democratic foundations of the United States, opposes the culture of free expression, and counters a tradition of unobstructed journalism and news-gathering in the country.

CHAPTER 4: THE SPECTER OF GOVERNMENT INTERVENTION

Chapter 4 addresses the specter of government intervention into the free flow of information through the European right to be forgotten. While the framers of the United States Constitution repudiated government regulation of freedom of speech and freedom of the press, the cornerstone principle of these tenets resides in the ability of an informed public to govern. With inherent "power of the people" to vet politicians, exercise a right to vote, and collaborate in social, economic, and political spheres, United States citizens balk at infringements of freedom of expression and any government movements resembling censorship. The enactment of the European right to be forgotten bestows government power on a private enterprise operating search engines and inculcates power into the hands of Google, a private corporation, to determine the availability of information in the global society. This phenomenon indicates that Google is operating in a quasi-governmental, regulatory capacity while it administers public rights on a global scale (Lee, 2016a). It has been suggested that implementation of the right to be forgotten should shift the adjudicatory responsibility from search engines to democratically accountable institutions (Stute, 2015). Today, Google and the other search engines sort through thousands of requests for de-linking.

While First Amendment opposition to the right to be forgotten dominates the landscape, other oppositional elements of government intervention remain, including the ability of the European Union to influence the free flow of information globally, thus compromising the availability of information for idea-sharing between citizens of Europe and the United States. Contemporary academic discussion focuses on the whether or not the search engine results merit First Amendment protection at all, with some (Volokh & Falk, 2012) arguing they should because the search engines are in fact speakers, even if an algorithm generates the speech. The notion of algorithms and generation of "machine speech" ushers a robust debate of the lines of demarcation between a speech act and a communication tool. Further, the role of the search engine is called into question: Is the search engine a simple conduit through which information passes or does the search engine itself exert editorial control over the information and change the composition of what we see by invoking or manipulating the algorithm? Since the right to be forgotten first emerged on the global stage, vocal objectors have come forth including Wikipedia founder Jimmy Wales, the British House of Lords, and a multitude of freedom of speech advocates from journalism and academia. Themes found in the resistance include the arguments that search engines can alter history, create memory holes, and alter the experience of citizens as they attempt to gain information by using a search engine. As the right to be forgotten undergoes acceptance among the 28 countries in the European Union with enforcement commencing in May 2018, the question arises whether or not a workable parallel may emerge in the United States. First Amendment scholars conclude that various parallels would not correspond in the United States. Yet, the notion of having information removed from search engines has received some attention, particularly though victims' rights organizations dealing with the aftermath of crimes, including the remedies for revenge pornography. U.S. law incorporates facets of forgiveness as found in bankruptcy and credit reporting. It has been suggested that the United States federal government should take the lead on the right to be forgotten, not jettisoning the rights to the states that could lead to a patchwork of various interpretations of data controller regulations (Bolton, 2015), with the Interstate Commerce Clause as a potential framework. However, it must be noted that the right itself is "built on shaky ground" because it relies on the premise of the search engine as a data controller, a designation that is different from the search engine role as a conduit or intermediary (Peguera, 2016).

Despite these considerations, the specter of government intervention into the free flow of information presents a real threat, allowing us to gain contemporary insights from philosopher Hannah Arendt, who addresses the emergence of a "lying world order" in totalitarian governments, promulgated by the obfuscation of truth that allows citizens to form opinions. The chapter concludes with a discussion of the absolutist theory of the First Amendment, precluding Congress from making a law to impede free speech in the United States.

CHAPTER 5: DISTORTING HISTORY: MEMORY HOLES AND OTHER DESTRUCTIVE FORCES

Chapter 5 tackles one of the most disastrous elements of the right to be forgotten—the fact that search engine de-linking leads to the distortion of history and the creation of memory holes. While proponents of the right to be forgotten point to its virtues for aiding privacy and allowing individuals to re-set the historical presentation of themselves online, the de-linking creates a form of censorship that can distort history and change the way individuals and societies perceive the sequence of events. The European Union has bestowed vast power on the search engines, private companies driven by financial motives, to police the availability of content and suggests a transfer of a government duty to a private corporation. So, the search engines now become the powerful "deciders" about whether information will appear or undergo removal once an individual has submitted a petition under the right to be forgotten. This line of argument assumes that the World Wide Web is in fact a public space, the web creates a public record, and Google, by this logic, therefore, becomes the indexer of personal records (Powles & Chaparro, 2015). Yet the search engine is now placed in a burdensome position to make decisions about the newsworthiness, timeliness, and impact of the information. The tiny adaptations of the historical record suggest a potential landslide of corrective information that would morph the understanding of events. The creation of these "memory holes" is reminiscent of George Orwell's dystopian tale, *1984*, detailing the acts of the Ministry of Truth as it tore memory holes in the society's memory to advance the prevailing political regime. Yet the power of rewriting history is not just found in works of fictional literature; it is also found in real world history, as both Adolph Hilter and Joseph Stalin rewrote historical documents and adapted history to advance the rise to political power and indoctrinate citizens with powerful totalitarian messages. While some would argue that taking a few links off the web will not re-write history, Western journalists and citizens must ask, "Where does it begin, and where does it go from here?" By examining the works of Hannah Arendt and contemplating the treachery of a falsified history, we may derive a contemporary perspective on the importance of a free flow of information in society. Through a brief discussion of the role of misinformation in a totalitarian society, we may find the warnings of philosophers such as Arendt prescient as the Digital Age introduces new vehicles for communication that can simultaneously inform and misinform.

The second topic tackled in this chapter encompasses the notion of memory studies, suggesting that with the ubiquity and acceptance of online search engine tools and digital archives, we are now moving into a societal state of "constant memory," quite different from the previous state of affairs that allowed for "forgetting," by virtue of the human mind's delicate memory. Now, however, we have the capacity to recall all, the sum total, suggesting the advent of a new digital archive fever, expounding on the notion first prescribed by Jacques Derrida in *Mal d'Archive*, anticipating the archival frenzy as a way to cope with death. For anyone attempting to understand the global impact of the right to be forgotten, the implications of Big Data storage and retrieval must be both philosophically and practically explored. Until we can navigate the confluence of legal, technical, and social dimensions resulting from the European right to be forgotten, we may not accurately determine the long-term effects on citizens living in the United States and the world. We will not soon forget the impact of the European General Data Protection Regulation that changes the structure and shape of information availability in Europe and beyond the European Union's borders.

CHAPTER 6: REFUTING THE PRIVACY ARGUMENT

This chapter addresses privacy, the prevailing concept that fuels the adoption and protection of the right to be forgotten in Europe. Privacy law has gained new power under the European General Data Protection Regulation, a power that is celebrated by some individuals and reviled by others. The chapter explores the dichotomy between the privacy perceptions of Europeans and U.S. citizens. Americans and Europeans seem to speak two different cultural languages when addressing privacy, with Americans underscoring "liberty" and "freedom from government intervention" while Europeans embrace notions of dignity, reputation, and honor. The right to be forgotten signals a call for greater protections for information privacy (Glazer, 2015), yet judicial precedence offers little protection for the right of individuals to prevent private information from being disseminated when the information is true. The various privacy interpretations arise not only from the contemporary practice, but also remain embedded in the cultural heritage, legal traditions, and historical frameworks for the founding of constitutions in the United States and Europe. In what has been called "irreconcilable" paths in the right to be forgotten, Americans bristle at the concept of removing data from the World Wide Web because it impairs the free flow of information in society. Meanwhile, privacy activists in the United States and Europe hold up the privacy movements as a welcomed and required protection in the Digital Age.

The chapter dissects the privacy argument and unveils a fallacy in presuming that the right to be forgotten addresses a privacy issue. While examining the history of privacy in America, the chapter addresses landmark cases that have proliferated the rights of working news media and have protected the free flow of information in society, even in the face of privacy claims. An analysis of ethics codes for the Society of Professional Journalists and the Public Relations Society of America suggest that right to be forgotten violates these ethics codes. Yet despite all of the opposition, Europe situates the right to be forgotten in a privacy context, giving data controllers both a power and a responsibility to regulate the free flow of information found on search engines. Europeans willingly look to the government to aid in privacy protection, while U.S. citizens seek freedom from government intervention and often view government regulation in privacy affairs as an offensive or aggressive action. Yet the Digital Age requires a greater appreciation for the new global audience that emerges online, the new-found permanence of information posted online and indexed by the search engines, and the nature of de-linking requests. The chapter further posits existing statutory regulation about privacy would suffice in the United States, a far better solution than data removal that creates memory holes and causes a distortion of history. In a three-pronged argument asserting that privacy is not the real issue at stake, this chapter points to the fact that government interference into the free flow of information is akin to government interference in one's home; existing privacy tort law may be utilized in the event of a privacy invasion; and, the information in question in the European right to be forgotten has already appeared in the public domain, making it not susceptible to privacy regulation by definition. Privacy activists, such as those affiliated with Consumer Watchdog, a national nonprofit, public-interest group, and First Amendment advocates, including academic scholars and legal analysts, remain locked in a tug of war, pulling on a rope tied to online governance, digital redemption, and the ability of private individuals to "forget" their past by de-linking data and reordering history. The chapter concludes by asking the reader to consider if U.S. citizens will tolerate the infringement of their rights to a cornucopia of online information for the sake of European privacy concerns.

CHAPTER 7: THE FREE SPEECH MANIFESTO FOR THE DIGITAL AGE: SEVEN TENETS TO PRESERVE INFORMATION INTEGRITY IN CONTEMPORARY DEMOCRACIES

Chapter 7 moves forward with a Free Speech Manifesto for the Digital Age: Seven Tenets to Preserve Information Integrity in Contemporary Democracies. This text suggests a "First Amendment Manifesto for the Digital Age," holding the world open to information exchange and giving citizens in the United States access to a marketplace of ideas that is not artificially tampered with through the erasure of information. The manifesto suggests principles including: (1) Uphold the First Amendment to preserve democracy. (2) Safeguard journalism with data availability. (3) Preserve historical memory. (4) Reject the conflation of privacy arguments with First Amendment protections. (5) Preserve the marketplace of ideas in the digital world. (6) Resist power shifts to private organizations with power over information availability. (7) Guard against global hegemony as the European Union exerts power over the availability of information found on the web. These tenets are proposed for contemplation of the role of the First Amendment in the Digital Age, the integration of freedom of information with democratic principles and the practice of democracy in the United States, and the recognition that encumbrances on information freedom suggest the inklings of totalitarianism.

CHAPTER 8: THE PREFERRED POSITION OF FREEDOM OF INFORMATION

With freedom of speech across the globe hanging in the balance, the text asserts that the First Amendment must be placed in a "preferred position" that protects freedom of information in the United States and abroad. In an examination of the preferred position doctrine, this chapter examines how the constitutional right to freedom of speech may supersede other rights claims in a court of law. The chapter addresses the newly emerged roles for the government, citizens, and private enterprise. Further, it confronts contemporary threats to freedom of speech and differentiates between the challenges associated with content regulation on the World Wide Web.

CHAPTER 9: DEFINING FREEDOM OF INFORMATION IN THE DIGITAL AGE

The book concludes with a discussion of information freedom in the Digital Age and defines areas that require future research and thoughtful consideration among scholars and practitioners devising online governance principles. The European General Data Protection Regulation provides a legal framework mandating Google's compliance to remove data upon user request if it meets certain requirements. Data once published can now be de-linked from the search engine. For now, though, in the United States citizens may still access information once published and not "cleansed" from the domains. Yet, the future remains uncertain. Today, Google functions as what some would call a "quasi-governmental agency," deciding what information to keep and what tidbits to erase. The company endures its own legal pressures including anti-trust lawsuits while functioning as a for-profit company.

This book dissects the "big pieces" of the impact of the European right to be forgotten from a sympathetic stance of the First Amendment, finding conflicts and compromises in freedom of information. The dynamic multimedia marketplaces and availability of information has changed the way citizens and consumers access news and information. The decisions made by governments, private corporations, and other influencers in the data storage and retention process present far-reaching implications for the media landscape.

Chapter 1

The Landscape

To understand the current context about the European right to be forgotten, we must explore the current landscape for web governance in Europe and across the globe. This book focuses on the provision in the European General Data Protection Regulation that allows for the removal of information from search engines upon approval of an individual's request and explains why United States media scholars reject the notion for interference with government. While global search engines struggle to find techniques to handle global governance, some activists proclaim that the "right to be forgotten" should be adopted universally across geographic borders to allow for the greatest protection of privacy rights. However, the prospect of implementing the right to be forgotten presents significant constitutional, ethical, and legal complications, in particular for U.S.-based media businesses and search engines. When contemplating the right to be forgotten as applied to search engines in the United States, a multitude of media law issues emerge. The argument contained in this text invokes the relevance of constitutional theories of the First Amendment to shine light on the fact that government regulation of data infringes on the free flow of information in a democratic society. This book asserts the right to be forgotten counters a commonly held belief in the "marketplace of ideas" and can impair the democratic process through inhibiting the unbridled flow of information in society by opting instead for an atmosphere of censorship. Advocates of the right to be forgotten claim the provision secures personal privacy in the Digital Age. This argument woefully veils the more troublesome issue of freedom of information in society. As an exploration of the legal and ethical conflicts that arise between the First Amendment and the right to be forgotten environments, we must first analyze the historical progression of the development of these laws and simultaneously consider the variable perspectives of individuals in these countries.

HISTORY OF THE FIRST AMENDMENT

Beyond the security guards and sheltered beneath protective glass, visitors can view the U.S. Bill of Rights in the Rotunda of the National Archives and Records Administration Building at 700 Pennsylvania Avenue in Washington, D.C. The document remains on display, providing a focal point for American citizens to celebrate the rights upon which the United States was built. While the first 10 amendments to the United States Constitution comprise the Bill of Rights, supplemental amendments have been made to account for a myriad of other designated rights, including the prohibition against slavery, term limits for the U.S. president, and succession of leadership in the event of the President's death. When it comes to the current conflict between the right to be forgotten and constitutional rights, it is important to examine the First Amendment of the United States:

> Congress shall make no law respecting an establishment of religion, or prohibiting the free exercise thereof; or abridging the freedom of speech, or of the press, or of the right of the people peaceably to assemble, and to petition the Government for a redress of grievances. (Amar, 2009)

Ratified in 1791, the First Amendment presents a virtual laundry list of precious rights, articulated because the memories of the Constitutional framers could not easily forget the revolutionary dimensions of independence from England and the oppression of Colonial America. The early adopters of the Bill of Rights could easily remember the techniques of licensing, taxes, and sedition laws used by England to restrict the free flow of information in Colonial America. While crafted as a Bill of Rights, the document presents a critical limit on government power. Credit for the Bill of Rights authorship goes to James Madison, who introduced the original version with 19 original amendments, subsequently whittled down by the U.S. House and Senate to 12 amendments for ratification by the states. The Bill of Rights was subsequently adopted as a bundle of 10 rights. Scholars indicate that Madison relied heavily on the Virginia Bill of Rights written by George Mason (National Constitution Center, 2017) and paid homage to the anti-federalist viewpoint that called for a document to protect individual human rights.

The Bill of Rights arrived after the adoption of the U.S. Constitution in 1789 and today, more than 200 years after ratification the First Amendment produces an ongoing, healthy dialogue in the United States about freedom of speech and freedom of the press. The First Amendment has weathered waves of contentions over sedition, commercial speech, and libel, and today the amendment encounters notions of government regulation, data retention, and Big Data analysis. While James Madison and his ratifying colleagues may not have considered the advent of the World Wide Web and the exchange of gigantic volumes of data each day, they provided U.S. citizens with a living document that allows contemporary interpretation and analysis to fathom the risks associated with implementation of government-sanctioned right to be forgotten protocols. Historically, the preliminary tests to the First Amendment dealt with the ability of citizens to speak out against the government, as evidenced by the Alien and Sedition Act of 1798, signed by John Adams. The act was intended to punish acts and statements against the United States government amid growing acrimony with France. The law stated that it was unlawful

> to write, print, utter or publish, or cause it to be done, or assist in it, any false, scandalous, and malicious writing against the government of the United States, or either House of Congress, or the President, with intent to defame, or bring either into contempt or disrepute, or to excite against either the hatred of the people of the United States, or to stir up sedition, or to excite unlawful combinations against the government, or to resist it, or to aid or encourage hostile designs of foreign nations. (*Constitution Society*, 2017)

While this was an historic challenge to the First Amendment, the ability for citizens to criticize the government remains a continuing debate point, particularly when considering activities during war or a national emergency, as arose in World War II. Today, the preservation of national security and the maintenance of state secrets remain as limits to freedom of expression.

As the author of the Bill of Rights, James Madison spoke out against the Alien and Sedition Act, by underscoring the rights of the people in the new democracy. In the Virginia Resolutions, written in 1799 and approved by the Virginia legislature in 1800, he wrote: "The People, not the Government, possess the absolute sovereignty. The Legislature, no less than the Executive, is under limitations of power. Encroachments are regarded as possible from the one as well as from the other. Hence, in the United States the great and essential rights of the people are secured against legislative as well as against executive ambition" (*Founders Constitution*, 2017). The First Amendment performs many functions protecting the nation by allowing for the transmission of information that ultimately helps citizens govern themselves in the democracy and keep a watchful eye on government power. While often cited as a catch-all for freedom of expression, it is important to note the First Amendment exclusively applies to state action, or in other words, the government regulation of free speech (Messenger, 2012). Therefore, we can extrapolate that First Amendment rights do not apply when censorship occurs from a private party. Many United States citizens have already agreed to a censorship environment when they accept the users agreement of a social media outlet. For example, Facebook, YouTube, and Instagram require users to comply with content regulations though terms of use (Facebook, 2017; YouTube, 2017; Instagram, 2017). By enrolling in the social media site, the individual agrees to abide by terms of use that give the user rights to use the channel, but at the same time governs the user's behaviors and acceptability of postings. Google overtly indicates the company's privacy policy in the user agreement (Google, Privacy Policy, 2017). In these cases, citizens purposefully accept permission for Google to regulate content in exchange for the use of the technology tool. As the World Wide Web heartily hums with global data exchanges, humans across the world willingly enter data into transactional websites and social media. Yet, under the European right to be forgotten, some citizens may at a later date make a claim to remove data in an effort to protect a personal reputation.

In the United States, government entities have not taken actions to compel compliance with the tenet of the right to be forgotten; however, countries in the European Union embrace the regulations. In practice, it is not the government making censorship decisions but rather a private company. In examination of the contemporary application of the European General Data Protection Regulation, we observe an interesting phenomenon: A private company executes the data removal, propelling the creation of memory holes with the affirmation of a government regulation. The search engine's removal of data represents the phenomenon of private companies responding to directions of a specific law, the European General Data Protection Regulation. The right to be forgotten in Europe is statutorily constructed with a legal heritage stemming from Regulation 95/46, originally passed in 1995. A corresponding statutory construction falls on shaky ground in the United States.

By learning more about the free flow of information in society, citizens can enlighten themselves. However, to the dismay of journalists and advocates of freedom of speech, general knowledge of the right to freedom of information is not ubiquitous. In the United States, 29 percent of 1,000 people surveyed in a study completed by the First Amendment Center could not name one right guaranteed by the First Amendment (First Amendment Center, 2017). For those who could name the rights, 68 percent mentioned freedom of speech, followed by freedom of religion at 29 percent (First Amendment Center, 2017). So, while the educated few may have learned about the First Amendment in a high school civics class or heard a mention in a pop culture movie, few understand the right that protects their freedom of expression against government intervention and preserves their rights as "listeners" to gain information in the marketplace. For U.S. citizens, the Supreme Court has recognized the right of citizens to receive handbills (*Martin v. Struthers*, 1943), to hear information from a union organizer (*Thomas v. Collins*, 1945), and to enable broadcasters to operate in the public interest and maintain responsibilities to listeners (*Red Lion Broadcasting Co. v. FCC*, 1969). Scholarship (Chandler, 2007) has asserted that theories of free speech are based on a fundamental premise that communication is a relationship between a speaker and a listener. The right to freedom of speech has also come to mean not only a right to express oneself, but also a right to access information (Oltmann, 2016), thus affirming the role of libraries in society. However, the presence of selection intermediaries like online search engines now block or bias the transmission of speech to listeners. It has been suggested that because of the dynamic nature of the communication process and technology innovation, the marketplace of ideas may need a legal affirmative "push" to preserve access. Through the creation of laws to preserve public availability to media for freedom of expression, a legal right preserving access would transcend the traditional marketplace forces (Barron, 1967). A stream of academic thought suggests that a new way to consider freedom of speech should be "free speech consequentialism," which contemplates the harm caused, thus measuring the benefits versus the maleficence.

Since its ratification in 1791, the First Amendment has sustained numerous disruptive media forces, including the advent of mass-produced newspapers, radio, television, satellite, cable, and online search engines. Despite these innovations, the tenets of the First Amendment have prevailed against government intervention. Today, some media professionals in the United States practicing journalism and public relations, who traditionally advocate for the free flow of information and disclosure, now question the potential impact of a similar legal action to the European General Data Protection Regulation in the United States. What if Google honors the right to be forgotten in the United States? What if citizens can formally request data removal, thus manipulating the digital archive available to citizens seeking information on the web? Consumers across the globe have evolved in recent years and have created a dependency on the availability of unencumbered online data and the search engines that retrieve the information. Therefore, media observers may claim the contemporary marketplace of ideas shift to the online venue is reflective of a convergence culture (Jenkins, 2006) that has adapted the relationship among technologies, media industries, and audiences, thus simultaneously altering the "logic" whereby consumers partake of media. Yet, whether our feet stand in the United States or in the European Union, we are collective citizens in the global convergence culture. This modern environment presents the circumstances in which the European Union called for the development of the right to be forgotten.

HISTORY OF THE RIGHT TO BE FORGOTTEN

At issue in this debate is the "right to be forgotten," a legal construct that has captured the attention of scholars and legal analysts in the United States and abroad. The notion of the right to be forgotten, which may be defined as third parties "forgetting" your past through the removal of data, presents a controversial topic among legal scholars, government officials, and private enterprises confronting the issue of government intervention into data control matters. The term "right to be forgotten" was proposed in the 2012 draft of the European General Data Protection Regulation in Article 17, stipulating that an individual should have the right to apply for data removal (Gilbert, 2015). The European terminology has encompassed various characterizations from "oblivion," which allows for the full deletion of certain public data, to the "right of erasure," which removes personal data at the subject's request, terms that lead to confusion among legal scholars about the precise definition and implications of the right to be forgotten (Ambrose & Ausloos, 2013). Today, the right to be forgotten remains a topic of concern in light of the European General Data Protection Regulation as search engines like Google are required to implement the regulation. The European Parliament and Council modernized the data protection rules in place since 1995 by incorporating the right to be forgotten and the right to notification when data has been hacked (European Commission news release, 2015). The moves followed numerous memoranda and public statements laying the legal grounding for the Act, including a May 2014 ruling by the Court of Justice of the European Union that found specified users may ask the search engines to remove results for queries involving the person's name. Google stipulates the results shown must be "inadequate, irrelevant, no longer relevant, or excessive" (Google, FAQ, 2017), setting forth qualitative measures that can be arbitrarily enforced. The administration of right to be forgotten provisions presents complications, particularly for media companies and search engines operating on U.S. soil. For opponents of both the regulation and the implementation of the right to be forgotten for search engines in the United States, a wealth of media law issues apply, notably constitutional arguments about the First Amendment. Scholars admit that the viability of a right to be forgotten that accommodates "take down" orders will not likely emerge in the United States because of the constitutional right to free speech (Sandefur, 2015).

Search engines worldwide face complex legal issues surrounding the implementation of the right to be forgotten. In February 2016, Google announced it would extend the "right to be forgotten" to all of the company's domains accessed in the European Union (Gibbs, 2016). In light of these movements, pressure in the United States continues to mount, as the European standard extends through Google's global reach. Decisions in Europe to adopt the right to be forgotten and subsequent cases have prompted U.S. privacy activists to take action and advocate for the similar implementation, calling the failure to do so in the United States an unfair business practice (Consumer Watchdog, 2015). The claim of an unfair business practice stems from allegations that Google allows for the right to be forgotten in Europe, but not in the United States. Activists have stated that Google should reflect on the fact they are not consistently treating people on both sides of the Atlantic (Simpson & Bates, 2015). Scholars have asserted that the inequitable management of data on either side of the Atlantic Ocean could create information "hot spots," but, if the search engines make information consistent across all domains, that means that the EU has created a "de facto" law in the United States (Owings, 2015). The thought of the European Union creating a law that ascribes far-reaching implications to freedom of information in the United States is intolerable for enthusiasts of the First Amendment. The prospective status of the right to be forgotten in the United States conjures a multitude of First Amendment issues that prompt debate about the role of privacy, legal intervention, and the retention of historical events. Further, the origin of the debate has deep historical roots, with the European notion of privacy derived from the concepts of dignity, honor, and personal respect, in contrast to the U.S. privacy tenets that rely on the notions of liberty and protection from state intervention (Whitman, 2004).

The current position for the right to be forgotten represents the result of legal actions in Europe. In May 2014, the Court of Justice of the European Union (CJEU) delivered a ruling in *Google Spain SL, Google Inc. v. Agencia Espanola de Proteccion de Datas* (AEPD), Mario Costeja Gonzalez. In the legal action, Costeja, a Spanish lawyer, sought removal of online information from a Catalonian newspaper and Google about a real estate sale in conjunction with bankruptcy proceedings under the presumption that the proceedings were resolved and no longer relevant. While the notices on the newspaper site were deemed lawful, the AEPD ordered the removal of the Google hyperlinks. This ruling has been credited with establishing the "right to be forgotten," and it gave search engine operators a broadened role for enforcement. The Costeja case thus called into question the policies and data activities of Google and other search engines worldwide. The right to be forgotten was incorporated into the proposal for a Data Protection Regulation (DP Regulation) in 2010, subsequently adopted in April 2016 as the European General Data Protection and will formally replace the 1995 Data Protection Regulation (DP Regulation) in May 2018.

Today, Google, a multinational company and the most frequently used worldwide search engine tool, is currently addressing the global issue surrounding the implementation of the right to be forgotten. In February 2016 after European criticism, Google announced it would extend the right to be forgotten to all of the company's localized domains accessed in the European Union (Gibbs, 2016). The contemplation of a U.S. right to be forgotten occurs in a multicultural environment, where differing values are placed on privacy and fundamental tenets for the free flow of information in society. European restrictions on data availability have repercussions beyond European borders.

The Right to Be Forgotten in Europe

Europe's adoption of the European General Data Protection Regulation translates into the implementation of the right to be forgotten in 28 countries across Europe. This regulation gives citizens the ability to request data removal about one's past—what some consider a human right—thus creating a legal mandate for Google, Bing, and other data controllers to remove data that is deemed inadequate, irrelevant, or no longer relevant. Since the onset of the right to be forgotten, Google has actively honored requests, facilitating multiple removals from the BBC, *The Daily Telegraph*, and other reputable media outlets. The BBC reports removal of stories dealing with a British Navy chief's "wrestling fetish" and the case of an optician who was ordered to pay reparations to a patient after laser surgery left her eyes overly sensitive to light (BBC, 2017). The *Telegraph* reports that the right to be forgotten has led to the removal of numerous hyperlinks, including the deletion of a story about a physician who attempted to drug his pregnant mistress in an effort to miscarry their baby, and an article about the conviction of a law student for killing and burying his father in concrete, because he urged his son to study at the Sorbonne in Paris rather than live with his London girlfriend (*Telegraph*, 2017a). The list goes on, showing the de-linking of information related to a call girl ring, corrupt sports referees, bank officials prosecuted for fraud, and investors imprisoned for stealing cash from an emerging company (*Telegraph*, 2017a). While the removal of these hyperlinks may provide temporary relief to the individual subject, the newsworthiness of these events may prompt pause. Wikipedia founder Jimmy Wales has called the right to be forgotten "deeply immoral" and chastised it for creating an Internet riddled with memory holes (*Telegraph*, 2017b). While speaking to the Wikimania conference in 2014, Wales stated: "History is a human right and one of the worst things a person can do is attempt to use force to silence another" (*Telegraph*, 2017b). In May 2015, 80 technology scholars sent an open letter to Google about the release of right to be forgotten data (Medium, 2017). The scholars requested aggregate data about Google's response to the right to be forgotten, indicating that it is important because Google and other search engines are now empowered to decide on an a la carte basis between personal privacy and retrievable information. The scholars stated, referring to the "right to be forgotten" with the acronym RTBF: "The vast majority of these decisions face no public scrutiny, though they shape public discourse. What's more, the values at work in this process will/should inform information policy around the world. A fact-free debate about the RTBF is in one's interest" (Medium, 2017).

In tandem with approval of the European General Data Protection Regulation, Google set up an advisory council for the right to be forgotten that issued a report February 6, 2015, setting criteria for de-listing request and indicating procedural elements (Google, The Advisory Council on the Right to be Forgotten, 2017). The report, published after the group conducted seven public meetings in Madrid, Rome, Paris, Warsaw, Berlin, Brussels, and London, uniting representatives from government, private industry and academia, details instructions for assessing the subject's role in public life, determining how the information is biased between the individual's privacy interest versus a public interest. The second half of the report details procedural elements, including recommendations about how to create an information de-linking request and how to challenge a de-linking request. The advisory council also set forth recommendations for transparency, calling for the routine reporting of aggregate data about de-linked information, including reports about the process and criteria used for de-linking. But the appendix offers the most interesting aspect of the report. Five of the eight individual council members published "dissenting opinions," advancing considerations about the global discussion of personal privacy and freedom of information. While one member called for the achievement of balance in these areas, U.S. entrepreneur Jimmy Wales takes a straightforward approach to explaining why he thinks the European right to be forgotten threatens the integrity of the free flow of information: "I completely oppose the legal situation in which a commercial company is forced to become the judge of our most fundamental rights of expression and privacy, without allowing any appropriate procedure for appeal by publishers whose works are being suppressed" (Google, The Advisory Council to the Right to be Forgotten, p. 27, 2017). Today, Google's website displays an annual transparency report, *European Privacy Requests for Search Removals*, which focuses on requests made on the European General Data Protection's Regulation for the European right to be forgotten, thus detailing the total number of requests made, the quantity honored, the countries where removals occur, examples of removals, and websites most frequently impacted by de-linking. The company also divulges information about the removal of hyperlinks, based on government requests and for copyright reasons. While this data reporting expresses an overture to accurate reporting attempts to address the concerns expressed in the open letter from 80 academics, further aggregation of data is needed to better discern the impact of the right to be forgotten de-linking in Europe.

For citizens of the European Union, the request process for search engine de-linking begins with webforms supplied by search engines in Europe. For example, the Google form offers a simple online request survey asking for the name used in the search, the name of the requestor, if that requestor is the individual or a representative such as a parent or attorney, the relevant email address, the URL for removal, rationale for the request, identity verification, a statement of accuracy authentication, and digital signature (Google, Search Removal Request under Data Protection Law in Europe, 2017). Another popular search engine, Bing, utilizes more questions in the preliminary form to discern the categorization of the individual as a private individual or one subject to the public eye of scrutiny. The Bing form, "Request to Block Bing Search Results in Europe," asks the requestor to enter the name, country of residence, upload identify verification, and provide an email address. Part of the form asks whether the individual is a public figure (politician, celebrity, etc.) and queries, "Do you have a role or expect to have a role in your local community or more broadly that involves leadership, trust or safety (for example teacher, clergy, community leader, policy, doctor, etc.)?" Then it asks the respondent to describe in detail their role in society, if they answered "yes" to one of the first two questions. The third part of the Bing form asks the individual to enter the Uniform Resource Locator (URL) that is requested for de-linking and gives four reasons to choose why the block is requested: inaccurate or false; incomplete or inadequate; out-of-date or no longer relevant; or excessive or otherwise inappropriate. The form then seeks supplemental evidence, such as a court order to verify the inaccuracy or falsity. The respondent then has the opportunity to add supplemental information when responding to the question, "Please provide any other relevant information describing your privacy interest in having this information blocked in response to searches on the name specified above and why you believe your privacy interests outweigh the public's interest in free expression and the availability of information." The individual must indicate if they have previously requested that Bing block the same page. In the fourth section of the form, the individual certifies that they are the individual named on the form, the URL in question relates to them and not to another person with a same or similar name, and the information provided is correct and accurate (Bing, 2017). While the de-linking process seems straightforward, data removal creates an information hole.

THE CONFLICT BETWEEN THE FIRST AMENDMENT AND THE RIGHT TO BE FORGOTTEN

In a full examination of the conflict between the First Amendment and the right to be forgotten, four key communication tenets serve as guiding principles for conducting a comparative critique that offers insights into the practice of and adherence to the First Amendment in the United States and the right to be forgotten in Europe. First, the implementation of the right to be forgotten in an era of Big Data assumes the formation of a grand arbiter who administers decisions about content removal, thus threatening the free flow information by placing the power of content judgments in a centralized, private source. Second, the arbitrary creation of a private "data controller" affronts the First Amendment by precluding the free flow of information, and endangers the robust debate that characterizes the nation and serves as a foundation for the nation's democracy. Third, the European right to be forgotten standard lacks guideposts, providing little definition about information that is no longer relevant (Cunningham, 2016). With the absence of standards or guidelines, the decisions are left to individuals making random decisions, individuals who lack formal training to accomplish their jobs. Fourth, an inherent conflict with freedom of expression erupts when search engines remove data. The right to be forgotten seeks to ensure privacy and relies on the right of the individual to "correct information," while the First Amendment offers freedom of speech. Despite the loud and convoluted cacophony of discussion about the issue's alleged constitutional conflicts, a solid issue remains: access to information. The average citizen in the United States and abroad may question how they will gain access to information if the government controls the availability. How can citizens arrive at a common understanding of truth, if transparency is routinely obliterated? What happens to society's history, if holes are ripped into the available inventory of retrievable information? Finally, can society endure the confusion that results from routine censorship? While we can consider the right to be forgotten in Europe, U.S. audiences bristle at the notion because it: (1) inserts government intervention into the free flow of information; (2) damages transparency by creating holes in the availability of retrievable data; (3) harms historical accuracy and clarity; and (4) endorses censorship, devaluing society's interest in valid political information. These consequences demonstrate inherent conflicts with the well-embraced First Amendment freedoms and principles.

Table 1.1: A Comparative Scorecard of Core Communication Tenets under the First Amendment and the Right to Be Forgotten Environments

Communication Tenets	*First Amendment*	*Right to Be Forgotten*
Information Flow	+	-
Transparency	+	-
Historical accuracy	+	-
Censorship	-	+

Government Intervention in Information Flow

Unencumbered free flow of information represents a long-held value in the United States and other democracies. The abilities for citizens to gain access to information about political candidates, assess economic situations, learn about social concerns, and collaborate with others in our society represent bedrock principles of American democracy. From the country's original founding to today, the ability for concerned parties to convene and talk about relevant issues remains protected from governmental prior restraint or intervention, except in circumstances that could imperil national security or threats present during wartime. Printers in the early New World colonies prevailed over multiple attempts at information suppression found through taxation and government bonds. Yet, these prior restraint efforts did not quell the early American voices, and, instead, their dialogues rose as courageous founders organized a new representative democratic government. By protecting access to information generated by the nation's earliest printing presses, the marketplace of ideas was open for business, with free dialogue and debate dominating the environment. Thus, almost 250 years later, when overtures to embrace the right to be forgotten sail across the pond from Europe, the implementation of a similar threat to freedoms in the United States raises concern about government intervention. For example, if Congress should enact legislation to install a right to be forgotten in the United States, that very act would conflict with the precise language of the First Amendment that states, "Congress shall make no law . . ." If a federal agency like the Federal Trade Commission required U.S. search engines to honor the law in the United States to remove "unfair business practices," as those alleged by U.S. privacy-based advocates, the federal agency would be enacting administrative law that would counter the First Amendment and it would find itself in the realm of unconstitutional actions. Multiple constraints impinge the maneuverability of a right to be forgotten in the United States, because the country's citizens value transparency, historical accuracy, and freedom from censorship.

Damages Transparency

The popularity of transparent communication has emerged in business books and leadership genres to endorse leaders who reveal the business models and rationale for their management decisions. In 2002, Sarbanes Oxley debuted to promote truthfulness through fair and full disclosure in financial reporting. Permeating the spheres of economic, social, and political influence, transparency has emerged as a core value of the new millennium. In the Digital Age, the ability to "see" through web search engines typifies not simply a daily pastime for people at work and at pleasure, but more importantly has become so integral to our way of being that failures to honor the transparent environment undergo strong critique. Consider political candidates who fail to conduct interviews with news media, or businesses caught concealing relevant financial data that would influence the investment-worthiness of their corporation. These modern-era concealments disrupt transparency. As we examine transgressions to imperil transparency online, we observe a violation of an ethical value of the Digital Age—preservation of transparency at all costs. By sustaining transparency, the digital society may operate on an assumption that every citizen can gain access to a grand buffet of information without ever leaving their home, simply by entering search terms into a recognized search engine. Today, at least in the United States and other non-European democracies, freedom of information advocates view the movement to greater transparency with search engine technology as a doorway to a panorama of accessible information and disdain any form of censorship that impairs the digital translucence of online search engines.

Harms Historical Accuracy

Citizens and journalists enter the marketplace of ideas to seek information about a variety of topics from political candidates and historical queries to social concerns and economics. While entering the marketplace, Internet users in the United States approach their keyboard with a psychological presumption that the information is "out there." If he or she just keep digging, the information will appear in the search results, and "voila" the individual will gain access to the sought-after data. This presumption, however, is based on a more fundamental tenet: The information is truly "out there" and may traverse through the databases of the online search engines to gain recall and ultimately render on the computer screen. In a right to be forgotten environment, the potential data recalled may have undergone "deletion" and therefore does not exist, vaporizing availability of formerly retrievable information.

A second presumption necessary for consideration when contemplating historical accuracy on the web is the phenomenon of "digital decay." Digital decay occurs daily as a side effect of the Information Age. As the Internet hums with data delivery and preservation, broken hyperlinks, reconfigured servers, replacement websites, inaccessible databases, and other technology-driven barriers facilitate digital decay. So why do we suffer angst about the right to be forgotten, if digital decay occurs anyway? An investigation into the critical differences reveals distinct variations, when we keenly examine the impairment that occurs in data retrieval under these two variable circumstances. An initial difference reveals content-specific intervention that occurs in the right to be forgotten in contrast to the breakdown of information that takes place more organically without specific intent to censor. A second difference indicates a presumed interference with the marketplace. The next chapter details the long-held analogy of the marketplace of ideas as a robust bazaar of ideas found in a healthy democratic state. The right to be forgotten inhibits the natural market forces that allow an idea to gain acceptance because information is removed. New York arts collective Rhizome (2017) has seized the moment as society addresses information storage and offers multiple programs for digital preservation of online content. By searching the Wayback Machine (https://archive.org/), visitors can capture a glimpse into the World Wide Web of yesteryear, perhaps taking a look at a defunct dotcom company or a website of a failed political candidate. Rhizome also provides a participatory environment that enables guests to upload recordings of contemporary webpages to ensure preservation.

Endorses Censorship

While privacy activists may proclaim the right to be forgotten as a victory for individual rights and privacy protection, others in society perceive the removal of information and Internet de-linking as censorship. To explore this debate more fully, we may consider Internet censorship as an aspect of control or suppression administered via a government or private enterprise that alters availability of information that can be accessed, published, or viewed on the Internet. Web censorship exists today in totalitarian states, and the right to be forgotten endorses censorship by allowing material to be extracted from the inventory of available information, based on an individual's claims that it is no longer valid. A study by The Internet Society, a U.S.-based non-profit organization that addresses Internet standards, access, and policy, found that nearly a third of Internet users strongly agree censorship exists on the Internet (*Global Internet User Survey*, 2012). The study, which surveyed 10,000 Internet users in 20 countries, revealed thought-provoking findings about the role of government in Internet regulation. Researchers found more than 70 percent of users agreed or strongly agreed that more government involvement would make the Internet too controlled or would limit accessible content. This finding suggests that the vast majority of Internet users recognize the specter of censorship and want to diminish it because it limits the ability to access information. Censorship impairs data availability, and for Internet users of the world, this presents a significant problem. The same report found that 66 percent of respondents agreed or strongly agreed that "censorship-free countries have a responsibility to keep the Internet free of censorship in countries where the Internet is being censored, controlled or shut down" (*Global Internet User Survey*, 2012). This polling result represents an inordinately high challenge, considering diplomacy and governing sovereignty, but denotes the pressing issues of government censorship should undergo healthy debate on the global stage.

In the current environment, controversy over the right to be forgotten spawns a discussion of the multicultural interpretations of privacy. However, the U.S. constitutional right of First Amendment freedoms tips the balance in favor of the free flow of information in society. U.S. citizens, beholden to values of liberty and protection from government intervention, vehemently argue the right to be forgotten precludes freedom of accessible information and can lead to information censorship in society. Some scholars fear the right to be forgotten could lead to corporate censorship, as Google may seek to minimize fines or legal actions and cautiously remove ambiguous cases (Simpson & Bates, 2015). This book addresses censorship in chapter 6.

Memory Holes

While the interruption to information flow and the obstruction to transparency prompt disgust and outrage from First Amendment advocates, the memory aspect of censorship presents a frontal attack on the free flow of information in society. Author George Orwell addresses the concepts of memory and censorship in the book *1984*, offering a glimpse into the manifestation of a fictional totalitarian world. Published in 1949, Orwell's work contains haunting dimensions of dictators, including those of Hitler, Mussolini, and Stalin, as he articulates a futuristic "purgatory" (Orwell, 1949). An analysis of the Orwellian text reveals "Big Brother" party slogans, featuring oxymorons such as "War Is Peace," "Freedom is Slavery," and "Ignorance is Truth." Further examination reveals Orwell's omniscience to address notions of censorship and memory in a totalitarian world. In particular, Orwell conceives of "memory holes," characterized as pneumatic tubes employed by the protagonist to deliver censored documents to a large incinerator and destroy remnants of the antagonist's past to Big Brother (Orwell, 1949). Readers of Orwell's book, often found on required reading lists for high school English classes in the United States, generate sympathy for protagonist Winston Smith, who attempts to navigate the surveillance state. The Thought Police found in *1984* obliterate Winston's own ramblings in a personal dairy. He writes, "to the future or to the past, to a time when thought is free, when men are different from one another and do not live alone—to a time when truth exists and what is done cannot be undone. . . . From the age of uniformity, from the age of solitude, from the Age of Big Brother, from the age of double-think—greetings" (Orwell 1949, p. 28).

While we stand decades beyond Orwell's prognostications about the year 1984, the lessons contained in the fictional text seem even more relevant today, as societies grapple with search engines, the right to be forgotten, and data retention. Let's take for example Winston Smith's lament about the absence of accurate records in the fictional dystopia: "When there were no external records that you could refer to, even the outline of your own life lost its sharpness" (Orwell, 1949, p. 32). With this literary text in mind, we can consider the right to be forgotten and the subsequent result to obliterate information in society, thus gagging free expression. With selective eradication of information from the web, will we ultimately wake up in a totalitarian dictatorship intolerant of the free flow of information, the expression of innovative ideas, and healthy political dialogues between individuals? While some may argue this is an apocalyptic prediction, poking holes in the collective memory of a society about a specific person or event will forever transform democracy and society's available body of knowledge.

Areas of Investigation

The implementation of the right to be forgotten in Europe foretells the treatment of data in the European Union and reveals crucial differences between the United States and European marketplaces about data privacy and information. Data controllers in Europe are legally required to execute the right to be forgotten, and face the daunting challenge of reviewing thousands of requests to remove information that individuals deem no longer relevant. While controversy over the right to be forgotten spawns a discussion of the multicultural understandings of privacy, the constitutional right of the First Amendment inherently tips the balance in favor of the free flow of information in society. U.S. citizens, beholden to values of liberty and protection from government intervention, vehemently argue the right to be forgotten precludes freedom of information and can lead to information censorship in society. Some scholars suggest vigilance to new technologies and respect for cultural differences are required to develop workable solutions for the right to be forgotten (Bennett, 2012; Ambrose, 2016). By setting privacy arguments aside and focusing on constitutional theories of the First Amendment, the argument against the right to be forgotten gains strength. The adherence to a free media environment that encourages public debate remains central to

American principles of democracy. The elasticity of the First Amendment continues to broaden as new modes of communication technology emerge, but it remains resolute to the freedom of expression principles woven into the cultural fabric of the United States. The effrontery to the inhibited flow of information, the implementation of censorship, and tolerance to government intervention contribute to a negative impact on democracy. The next chapter addresses the theoretical framework of the marketplace of ideas to further enlighten the inherent conflicts between the right to be forgotten and the free flow of information in society.

Chapter 2

The Marketplace of Ideas

The Disruption of Dynamics

Imagine walking into a fruit-and-vegetable market on a sunny summer morning. The merchants present their wares from colorful tables, showcasing the outcome of months of hard work, planting seeds, watering the plants, eliminating weeds, and carefully tending to the burgeoning buds to promote growth. Each farmer "competes" with the other to provide the most tantalizing strawberry or appealing seedless watermelon. The marketplace thrives, as customers arrive at the stands and fill shopping bags with the freshest fruits and vegetables to stock home pantries and create delicious meals. Now, let's envision the marketplace of ideas, as found in the robust exchange of information in our global society. Instead of fruits and vegetables, journalists present notions of social, economic, and political significance from their "carts," which are composed of print news articles, online news articles, and television news media segments, all presented by leading news media organizations. Some journalistic "farmers" adhere to professional ethics codes and subscribe to tenets of accuracy, objectivity, and a genuine concern for accuracy. However, in the marketplace, some unethical purveyors of media information disavow truth and accuracy, offering words and concepts that masquerade as news and accurate, legitimate information. To the unsophisticated consumer, a news article produced with objectivity and crafted with journalistic integrity may look the same as one fabricated by a citizen journalist or a pay-for-play content creator and published on a seemingly legitimate social media site. As visitors to the marketplace of ideas, citizens must then question the differences between these competing ideas, determining whether the news is real or fake. The active role of the audience presents a natural corollary to these thoughts, as individuals perceive the information, share stories, and virally communicate in social media environments. Citizens actively engage in the marketplace to choose the "best ideas," allowing the marketplace to appropriately perform.

To effectively contemplate the marketplace of ideas, let's define the marketplace itself—the philosophical origins, its maturation, and potential corruption of marketplace principles. Today, the marketplace faces a multitude of threats, notably the European right to be forgotten and the degradation of transparent news-gathering on a global basis. Fake news, "alternative facts" and incorrect information masqueraded as truth plague the current information environment. The evolving media environment and impact on marketplace of ideas dynamics require a full examination, as does the subsequent impact of the marketplace's demise on the public's perceptions of social, economic, and political structures. In the United States the First Amendment staunchly defends the free flow of information. However, the information marketplace is changing with the core tenets of truth, accuracy, and objectivity no longer deemed as valuable as they once were. These forces influence our society today and catalyze a deleterious effect on the ability of United States citizens to form a public opinion on relevant issues. The marketplace fragmentation promotes a confused electorate, composed of voters handicapped in their ability to discern truth from the crowded marketplace. Movements like the European right to be forgotten perform acts of censorship that obstruct the free flow of information in society and simultaneously controvert the First Amendment in the U.S. Constitution's Bill of Rights. In a right to be forgotten environment, authenticity and transparency suffer. The power to censor becomes the government-blessed right of a chosen few, forsaking natural marketplace dynamics. This compromised circumstance leads to an obfuscation or "warped" truth. The power equation morphs, and, in a censored digital information environment, erasure transforms the open marketplace into a covert, hidden repository of information not accessible by all citizens.

THE MARKETPLACE OF IDEAS DEFINED

So what is the marketplace of ideas? Why do we commonly apply marketplace principles to communication, information, and news in our society? This notion captivates philosophers, jurists, and First Amendment scholars. Characterizing the exchange of ideas in the context of a robust marketplace is not a new or novel concept, but rather reflects a documentable heritage composed of intersections with economic theory and constitutional law. The Marketplace of Ideas Theory, with origins in the works of John Milton and John Stuart Mill, was articulated by U.S. Supreme Court Justice Oliver Wendell Holmes, Jr., in 1919 in a U.S. Supreme Court dissenting opinion in *Abrams v. United States*, which stated "ultimate good desired is better reached by trade in ideas—the best test of truth is power of the thought to get itself accepted in the competition of the market, and that truth is the only ground upon which their wishes safely can be carried out. That at any rate is the theory of our Constitution" (*Abrams v. U.S.*, 1919). The ubiquity of the Internet enables global communications channels, linking people and corporations across miles, thus broadening the marketplace to encompass ideas created by both journalists and citizens. Closely tied to principles of market economics, the Marketplace of Ideas Theory resonates throughout American culture. Scholars in the United States readily apply the capitalistic concepts found in the marketplace of ideas to discussions of emerging news media models, the information society, and legal structures. Today, we extend the international information marketplace to the global village, as world citizens function in connected societies united by information networks, wireless technology, and constant connectivity.

HISTORY OF MARKETPLACE OF IDEAS

The marketplace of ideas has functioned as a cherished exchange in the formation of governments and civil structures. In fact, it is so embedded into thinking of the First Amendment that it resides as a bedrock principle for dialogue on government, civil, and social structures. To understand the evolution of the marketplace of ideas, we may look back to original tenets set forth by John Milton in *Areopagitica* (Milton, 1644), who called for freedom of expression in 1644 with a resounding repudiation of government censorship embedded in the Licensing Order Act of 1643. Milton suggested restriction of speech was not necessary because in a society guided by freedom and openness, truth would prevail. In a condemnation of censorship and licensing, Milton said that destruction of a book "kills reason itself and kills the image of God, as it were in his eyes" (Milton, 1644, p. 8). In a repudiation of the licensing of printing, Milton states that if society must regulate printing to remedy manners, the government should regulate all recreation of man's delight, including dancing and music (Milton, 1644, p. 21). While licensing of the press, taxation, and prior restraint represent techniques utilized by the British crown in an effort to quell the free flow of information in the New World, radicals of the American Revolution sought to preserve freedom of speech in the burgeoning United States of America, as evidenced by the First Amendment of the Bill of Rights.

Thomas Jefferson, third president of the United States, specifically itemized the diffusion of information and freedom of the press as hallmarks of liberty in the nascent nation, in line with other rights such as freedom of religion, thus forming what he characterized as a "bright constellation" that guided steps through revolution and reformation (Jefferson, 1801). He broadens the notion by stating that a departure from these ideals signals a road away from liberty:

> The wisdom of our sages and the blood of our heroes have been devoted to their attainment. They should be the creed of our political faith, the text of civic instruction, the touchstone by which to try the service of those we trust; and should we wander from them in moments of error or of alarm, let us hasten to retrace our steps and to regain the road which alone leads to peace, liberty, and safety. (Jefferson, First Inaugural Address, 1801, p. 10)

The steadfast link between speech and press freedoms and the notion of liberty reverberate through U.S. history. These same concepts subsequently emerged in John Stuart Mill's political writings in *On Liberty* (1859). He eloquently asserts liberty of thought cannot rightfully be separated from the cognate liberties of speaking and writing (Mill, 1859), as he incorporates concepts of freedom of speech and freedom of thought into the chapter titled, "Of the Liberty of Thought and Discussion":

> But the peculiar evil of silencing the expression of an opinion is, that it is robbing the human race, posterity as well as the existing generation; those who dissent from the opinion, still more than those who hold it. If the opinion is right, they are deprived of the opportunity of exchanging error for truth; if wrong, they lose, what is almost as great a benefit, the clearer perception and livelier impression of truth, produced by its collision with error. (Mill, 1859, p. 14)

This approach to freedom of expression indicates that citizens may arrive at truth examining opinions, casting aside those deemed erroneous in exchange for others that lead to a discernment of accuracy and truth in the society. Mill further underscores the importance of freedom of expression by repudiating censorship. He states we cannot operate with certainty that the opinions to be stifled are false, and, even if they were stifling them, the act of stifling still presents an evil (Mill, 1859, p. 14). The threads of freedom of speech, freedom of the press, and the ability to preserve and protect an open forum for debate of political, social, and economic issues were woven into the fabric of the U.S. Constitution and the culture of the New World. Activists rely on the presumed openness of the marketplace to exert innovative and revolutionary ideas. The contemporary environment suggests that we may require a rallying cry to keep the marketplace of ideas open.

In keeping with marketplace analogy, jurists have later applied these tenets to U.S. Supreme Court opinions, propelling the notions into the law of the land and making the metaphor inseparable from freedom of information.

The marketplace of ideas concepts later found its way to the Supreme Court of the United States, first cited by Justice Oliver Wendell Holmes in his famous dissent in *Abrams v. United States* (1919), when he discussed "free trade in ideas" within competition in the market. This statement, regarded among historians as a profound victory for freedom of speech, represented a shift from Holmes to a more progressive view of the First Amendment, moving away from the thinking that prior restraint was the only repugnant form of censorship (Healy, 2013). In the concurring opinion of *United States v. Rumely* (1953), Justice William O. Douglas states that publishers "bid for the minds of men in the marketplace." Lastly, the U.S. Supreme Court's ruling in *Brandenburg v. Ohio* (1969), elevated the marketplace of ideas notion as the prevailing school of thought to protect free speech, unless that speech promoted "imminent lawless action." To further examine the progression of the marketplace of ideas as a prevailing concept in freedom of speech protections, several U.S. Supreme Court cases provide points on this trajectory. On August 22, 1918, a group of five Russian immigrants threw leaflets from a New York City window, denounced World War I, and chastised U.S. efforts to impede the Russian Revolution. The defendants were charged with violation of the Sedition Act of 1918, an amendment to the Espionage Act of 1917. The Supreme Court held the criticism was not protected by the First Amendment because of advocacy for overthrowing the government. Justice Oliver Wendell Holmes did not agree with this decision and authored the dissent:

> But when men have realized that time has upset many fighting faiths, they may come to believe even more than they believe the very foundations of their own conduct that the ultimate good desired is better reached by free trade in ideas—that the best test of truth is the power of the thought to get itself accepted in the competition of the market, and that the truth is the only ground upon which their wishes safely may be carried out. That, at any rate, is the theory of our Constitution. (*Abrams v. United States*, 1919)

In a broad defense of free speech, Holmes asserts, "We should be eternally vigilant against attempts to check the expression of opinions that we loathe and believe to be fraught with death, unless they so imminently threaten immediate interference with the lawful and pressing purposes of the law that an immediate check is required to save the country" (*Abrams v. United States*, 1919). This decision occurred just months after another case involving the Espionage Act, *Schenck v. United States*. Just nine months earlier Holmes undertook a different legal approach, finding that leaflets designed to obstruct enlistment of soldiers violated the Espionage Act, and standing by the restriction on free speech in the event of a "clear and present danger," noting proximity and degree of the speech (*Schenck v. United States*, 1919). So what made him change his mind? Letters uncovered between Holmes and his contemporaries in the summer of 1919 reveal a change from a core belief that prior restraint was a clear violation of the First Amendment, but once speech is "spoken" and revealed into the marketplace, it could be found in violation of a multitude of laws (Healy, 2013). The marketplace of ideas had begun to take hold in Supreme Court rationale. By 1953, when the court was deciding the fate of a political book publisher who functioned as a lobbyist of sorts, Justice Douglas's concurring opinion states:

> We have here a publisher who, through books and pamphlets, seeks to reach the minds and hearts of the American people. He is different in some respects from other publishers. But the differences are minor. Like the publishers of newspapers, magazines, or books, this publisher bids for the minds of men in the market place of ideas. (*United States v. Rumely*, 1953)

Douglas goes further to state that freedom of speech and the press occupies a "preferred position" and that in the United States, there is "confidence that the safety of society depends on the tolerance of government for hostile, as well as friendly, criticism, that, in a community where men's minds are free, there must be room for the unorthodox, as well as orthodox views (*United States v. Rumely*, 1953). By 1969 when the Supreme Court encountered the case of Ku Klux Klan leader Clarence Brandenburg convicted of violating the Ohio State Criminal Syndicalism Act, the tide was turning toward testing action versus words. Moving beyond the "clear and present danger" test found in Schenck, the court moved to a new test, protecting speech unless it would lead to "imminent lawless action" (*Brandenburg v. Ohio*, 1969). In his concurring opinion, Justice Douglas cast aside the "clear and present danger test," referring to Justice Holmes's dissent in *Gitlow v. New York* (1925) that indicated ideas should flow freely: "in the long run, the beliefs expressed in proletarian dictatorship are destined to be accepted by the dominant forces of the community, the only meaning of free speech is that they should be given their chance and have their way" (*Gitlow v. New York*, 1925). While the court traversed a wandering road influenced by two World Wars and the Vietnam era, justices marked distinctions between words and action, yet remained loyal to the notion of freedom of speech preferred by James Madison and the founding fathers. In 1971 the Espionage Act was scrutinized again by the U.S. Supreme Court, as the justices ruled in *New York Times Co. v. United States* (403 U.S. 713) that the government could not exercise prior restraint to quell the publishing of government reports on the Vietnam War.

The marketplace's theoretical frame has traditionally provided a guiding force for journalism and the free flow of information in society. Moving beyond the borders of the United States, where the Constitution's Bill of Rights stands, other democratic governmental structures have applied similar protections of speech in the global environment, and citizens view these rights as valuable. Living and working in the Information Age, citizens report a theoretical adherence to freedom of information, whether or not it may be exercised in respective countries remains an altogether different question. Global attitude research of citizens living in 38 different countries, including the United States and countries in Latin America, Europe, Asia, Africa and the Middle East, found that more than half characterized the right to freedom of information as very important (Pew, 2015). To measure global support of free expression, Pew developed a Free Expression index, which ranks the United States, Canada, Poland, Spain, and the United Kingdom in the top five spots (Pew, 2015). In general, publics worldwide endorse freedom of speech and the press, though differences erupt when considering critique of the government and expression that is offensive to minorities or religious groups. While residents in the United States and Canada report a tolerance for criticism of the government, few believe people should be able to say things that are sexually explicit or that imperil national security (Pew 2015). Just as merchants compete for customers at the weekend fruit-and-vegetable market, activists, politicians, and advocates for social change compete for the minds of citizens of the world to solidify a consensus, thus allowing collective truth and understanding of the world in which we live to prevail. Yet, considering the contemporary news media environment, it is time to take pause and truly analyze whether or not a marketplace analogy accurately applies to unencumbered free flow of information.

CRITICISM OF THE MARKETPLACE OF IDEAS

Despite the well-embraced popularity of the marketplace of ideas, some scholars criticize the concept by dismissing it as a faulty metaphor designed to rationalize anti-regulation viewpoints (Joo, 2014). The marketplace of ideas concept itself squarely rests on the principle that markets themselves should operate in an unharnessed and unregulated fashion, one that can sustain criticism for its underlying assumption—a premise biased to capitalistic means. It has also been asserted individuals move in a "herd" and do not think for themselves. The marketplace of ideas analogy has been questioned for the presumption that people are rational consumers of information (Gordon, 1997), and that humans are susceptible to moving in a herd whereby they do not think for themselves. But rather, humans remain susceptible to different social pressures, intellectual concerns, and motivations not easily quantified. Further, the marketplace analogy implies that the best ideas will prevail. An alternative point of view on the marketplace suggests that prevailing ideas are the popular ones, or those held by the "majority," leading to a tyranny of the majority in defining what we think about and accept with little care for truth and accuracy (Gordon, 1997). The constant agony to find truth emerges from the marketplace's dynamics. Some scholars assert the thought that, while truth can undergo persecution, it cannot be suppressed (Mill, 1859). As humans wrestle to discern truth, the capability to differentiate fact from fiction becomes more vitally important.

Overall, the marketplace of ideas operates with broad presumptions. The marketplace assumes transparency for the presentation of concepts; accepts that data is presented with accuracy; and functions on the premise that humans interpret information with rationality of thought. Thus, whether citizens subscribe to the marketplace principles or question them, movements to censor the marketplace fly in the face of the market's exchange. Today, other issues conflate the marketplace and influence regulation of the Internet-based communication marketplace. While this book deals with the right to be forgotten and corresponding censorship, other issues such as the notions of net neutrality, Big Data manipulation, and information retention inform this discussion. The current era of Big Data unearths a myriad of issues, including information and analysis, and at first glance some legal scholars posit the current environment of Big Data can simultaneously allow for government *and* First Amendment protections (Ghosh, 2012). Some suggest that when considering personal data, we should consider it as a "trade secret," ascribing it certain legal rights (Malgieri, 2016). Yet, others have suggested the United States needs to adopt a free speech strategy that recognizes the right to personal information is not absolute and diminishes the contention that limitations to free speech represent a "slippery slope" (Kamatali, 2016). While these scholars attempt to make sense of the convoluted issues in Internet regulation and the restraints of information, from a U.S.-centric perspective, safeguarded by the constitutional protections of the First Amendment, this test asserts the right to be forgotten violates the concept of the marketplace of ideas.

HOW THE RIGHT TO BE FORGOTTEN VIOLATES THE MARKETPLACE

In theory, the marketplace of ideas functions appropriately without interference from marketplace dynamics. By looking to traditional economic principles, a marketplace may undergo disruption when government intervenes to set prices or support production. Manipulation by a third party changes the market's vitality. Prices rise, products are no longer available, and consumers settle for inferior products. By applying principles of market economics to the marketplace of ideas, we can observe how the right to be forgotten imperils the marketplace. First, the establishment of the right to be forgotten contradicts the concept of the marketplace of ideas and stifles the free flow of information in our society. The marketplace thus loses an inventory of information at the arbitrary hands of a search engine that liquidates data based on citizen requests. The power equation tips to the data controller. In essence, the data controller exercises prior restraint on the flow of information. The information, once freely available in the marketplace, is now "restrained" and removed. If the government agencies in the United States were to allow Google to remove data, based on the request of individuals, Google would be positioned to become a grand arbiter of information and thus might use its monopoly on permissible ideas to approve or deny access to the marketplace for specific social, political, and economic ideas.

For those concerned with protecting the marketplace of ideas, the arbitrary nature of removal requests spells trouble. In requests, citizens may seek to eradicate information that they may ostensibly deem no longer relevant, but, in reality, they may simply attempt to alter perceptions against poor taste, a bad decision, or seek to re-write their own personal history with the benefit of hindsight. As Google and other search engines implement the right to be forgotten in Europe, the process itself remains awkward and arbitrary, handled on a case-by-case basis. Google states the process begins when a web form initiated by an individual is submitted and is then reviewed on an individual basis, considering the information's accuracy, timeliness, and public interest (Google, Frequently Asked Questions, *Transparency Report,* 2017). The right to be forgotten opposes the marketplace of ideas theory, because it diminishes the ability for people to participate in the marketplace, it disrupts the natural process of communication, and it may misappropriate the marketplace regulation to a corporate procedure or legislative function, rather than to marketplace participants (Larson, 2013). Scholars examining the European General Data Protection Regulation in its nascent stages acknowledged the role of information intermediaries should not "be underestimated" (Tamo & George, 2014), suggesting that the search engines exercise an authority over the de-indexing process and implying that this authority could be abused. The robust marketplace of ideas encounters trouble when information is lost and the currency of accuracy is no longer valued. Information gaps formed by data removal generate dangerous forms of censorship that can influence democracy. The accurate re-telling of information becomes suspect, because gaps are created. The right to be forgotten shuts down information transactions, obliterates accuracy, and imperils freedoms inherent in the marketplace of ideas. Under the right to be forgotten, the power to control information rests with untrained "data controllers" like Google and other search engines, which determine the availability of information in the Digital Age. Scholars indicate that, when a de-listing request arrives at the data controller, a challenging, tripartite balance must be struck between the right to freedom of expression, the right to privacy, and the right to data protection (Kulk & Borgesius, 2014). It is during these censorship decisions that recorded truth can be altered, based on the de-linking decision by the search engine. Search engines as the sole arbiters of information remain a central sticking point for the right to be forgotten. While the European Union has anointed Google and other search engines as "data controllers" for society at large, the right to be forgotten represents the tip of the iceberg. Search engines will continue to gather data, perform data analytics, and utilize information for commercial purposes. Therefore, as participants in the Information Age, we now must come to recognize that information about us and procured by us during every search engine request remains catalogued and readily commercialized. A private entity controls this information, rather than the government, that manages our vital records such as marriages, births, and other personal information. Yet, search engines are capitalistic organizations driven by the market need to deliver profits to shareholders. So, do government entities have a legitimate right to interfere with the business practices of these private companies? This certainly requires further exploration, because government entities cope with citizen advocacy requests to make the right to be forgotten uniform across Europe and the United States. This book asserts the right to be forgotten conflicts with the First Amendment's intent by compromising the continuity of accurate information instead of respecting the free flow of information. Questions remain in this evolving circumstance. How do we navigate the right to information? What is the individual's legal and ethical right in light of the convergence with technology, the free flow of information, and the assertion of privacy rights?

Historical Gaps

The integrity of the communication process suffers irreparable harm when information is taken away from the search engine, creating a gap in history. While the right to be forgotten has been hailed by privacy activists, the European General Data Protection Regulation's right to be forgotten provisions remain chilling to First Amendment advocates in the United States, who assert that the right to be forgotten alters the historical record and makes information that was once lawfully public no longer available. This compromises an essential right to access to information (Manjoo, 2015). As further support of this argument, an examination of Article 19 of the United Nations Declaration asserts, "Everyone has the right to freedom of opinion and expression; this right includes freedom to hold opinions without interference and to seek, receive, and impart information and ideas through any media and regardless of frontiers" (United Nations, 2016). The declaration's language remains relevant to the current technology-driven information society and underscores the fundamental human right to have ideas shared through search engines as any media. Without the ability to accurately preserve and sustain information, some question the ability of democracy to appropriately function. Scholars (Aiello & Proffitt, 2008) have argued that citizens in a functioning democracy must have access to information to make "reasoned, rational decisions." In the functioning democracy news media play a vital role to vet political candidates, report on their past, and determine if they are appropriate representatives for the democracy. Therefore, any attempts to compromise this data integrity would inevitably obstruct the vetting process for candidates in a democracy. Let's consider how the right to be forgotten can encumber the media. In particular, the right to be forgotten threatens the power of the news media and informed citizenry to scrutinize a formerly private individual who has become a public figure by entering public political life for the first time and sacrifices the ability to identify emerging patterns of behavior or traits (Wechsler, 2015). In this system, the audience takes an active role—questioning authority, seeking additional information, and forming relationships with others. Yet, if the information is no longer accessible, the decision-making process declines in the absence of data. Thus, it may be argued the incorporation of the right to be forgotten imperils freedom and democracy, because media and everyday citizens no longer have access to an accurate recollection of facts and occurrences in public life.

Power Shifts

Who communicates, the content conveyed, and the total information inventory suffer compromise in a right to be forgotten environment. Transparency becomes murky with information now tattered by holes and blinded by gaps. Significantly, the marketplace dynamics are disrupted, and the information decision-maker becomes a single individual, not the robust marketplace. Thus, we observe a centralization of power into one individual, a stark contrast to the marketplace dynamics. In what have been coined as "deciders" (Rosen, 2012), the individuals at Google and other search engines stand to inherit great power of information management. This precarious/dangerous circumstance can then lead to an obfuscation of information, centralization of power over information dissemination, and the warping of "truth." Obfuscation contradicts the journalistic tenets of honest information-gathering and leads to a world where the power-brokers dictate the nature of truth. The product of once uninhibited information is gone from the marketplace, leaving space for

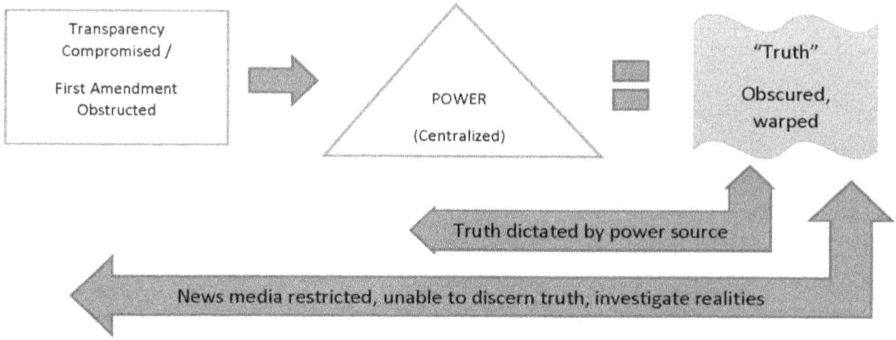

Figure 2.1: Dynamic Disruption in the Information Marketplace.
***Source*: Courtesy of the author.**

other ideas that "morph" to fill the void. These ideas may be dictated by the centralized power source and may ultimately lead to abuses of power.

GLOBAL TRANSPARENCY RIDDLED IN A RIGHT TO BE FORGOTTEN ENVIRONMENT

News-gatherers and citizens of the world may fear the punctures in global information transparency. While some may dismiss the European right to be forgotten as a concern for the European Union, the phenomenon influences the free flow of information beyond the borders of Europe. The right to be forgotten disrupts the transparent information marketplace, inculcating search engines with power over information curation and dissemination in Europe, thus changing the availability of data. Google and other search engines in Europe and other parts of the world become grand arbiters or "merchants" of information in this environment. Individuals make solitary decisions on whether to keep or remove information, wielding tremendous influence over the words and images that citizens may view in the marketplace. Each year Google publishes a Transparency Report that itemizes the unique record locator (URL) removal requests. According to the Google Transparency Report (2018) between 2014 and March 2018, the company had received 655,561 requests to delist with requests for de-linking at 2.4 million (GoogleTransparency Report, 2018). According to the report, France's annuaire.118712.fr and www.Facebook.com had the top number of URLs delisted, with 7,712 and 6,682, respectively (Google Transparency Report, 2018).

Citizen Impact

Studies have found that search engines provide a strong information acquisition tool to significantly contribute to political knowledge (Stephens et al., 2014). By placing the free flow of information in jeopardy because of information de-linking, the ability of citizens to evaluate the centers of power stands in question. The notion of rational decision-making may be altered by the creation of memory holes in the search engines. In the contemporary media-saturated society, the advent of "monitorial citizens" have been observed (Schudson, 1998), as citizens review online content and utilize all technology-enabled tools to scrutinize public issues (Feenstra, 2014). These are citizens who now perform a watchdog role on institutions of power by scouring information and using personal influence to communicate with others. The right to be forgotten cripples the citizen watchdogs, because the information revealed on the search engines no longer depicts a complete picture of the reported reality. In a right to be forgotten atmosphere, search engines may still exist, but they will not be functioning as the valuable, accurate tool they once were. Researchers suggest the search engine performs in a more precise manner than social media or news apps for acquisition of political knowledge, because citizens utilize social media to stay connected, and search engines perform a different role that requires greater user engagement to look for specific news information (Stephens, et al., 2014). Using a primary search engine for information, citizens often assume the information is accurate and accessible, taking for granted that the full scope of information available on a topic will render itself on the search engine results. Globally, citizens of 38 nations polled seem steadfast in the belief that it is very important to live in a country that provides Internet without government intervention (Pew, 2015). Yet, a right to be forgotten environment presumes that the information retrieved will contain "holes," from information de-linked at the request of the individual.

While search engines enjoy greater power to keep or remove information, the audience's role changes, and journalists must advance beyond search engine results to a more cumbersome technique for documenting information. Just as a consumer eager to buy a rare commodity pursues alternative means of procuring the item, including thievery or dealing with dishonest merchants, curious information netizens may cultivate subversive habits of gathering, such as information hacking by burrowing into the data-retention servers of information search engines. The rise of Anonymous and other citizen hacking organizations reveals the subversive desire to gain information access. Therefore, the audience's active role may emerge as an illegal quest for information. Further, as information consumers, the "customer" may experience a sense of loss or mourning over the once-robust marketplace, now reduced to a government-controlled, monopolistic atmosphere. The marketplace now resounds with a potentially depressive belief that the robust exchange will never re-emerge, leading to subversion and formation of alternative marketplaces.

Impairing Journalists and News-gatherers

Journalists are people too, equipped with smartphone technology, laptops, and tablet devices readily at hand. When web search engines honor requests to remove information, journalists experience a heavier burden, because data is not immediately and reliably available through a few keystrokes, but rather requires "old economy" ways of information-gathering, now perceived as hindrances. Journalists may have to return to courthouses and wade through hard-copy archive materials to uncover a story. Today, journalists exercise a research "shorthand," using all available technology to find sources, gain information, and target interview subjects. For the time being, adherence to the First Amendment may stem the tide against right to be forgotten censorship in the United States, but this does not protect information that is de-linked outside the United States and thus global information exchanges will continuously transform as data controllers remove information, new bytes of data emerge, and the confounded audiences struggle to find truth and accuracy in online information. Movements by the European Union to adopt right to be forgotten measures imperil the free flow of information in exchange for tenets of "erasure," thus placing a priority on personal privacy over the uninhibited free flow of information in a global society. While we address personal privacy arguments in chapter 5 and identify the conflict points, the compromise of the information marketplace leads to deleterious effects on both the individual citizen and that person as a member in broader society engaged in social, economic, and political discourse.

Damaged Marketplace Leads to Obstructed Transparency

To appreciate the multifaceted implications of the right to be forgotten on the marketplace of ideas, the fruit-and-vegetable stand analogy once again proves useful. Let's consider a hypothetical situation in which a government entity started to fix the prices of the fruits by incentivizing the farmers to produce only one variety of a designated fruit. With fewer choices offered, the product quality may decline, frustrating customers seeking a wider variety of tasty consumables. Industrious shoppers may travel to a nearby stand a few miles away because it is free from government intervention. Extending this notion to the free flow of information in a marketplace of ideas, government-endorsed censorship constricts the information exchange. Further, the forces of power shift away from the marketplace itself to a centralized information source empowered by the government. In the case of the right to be forgotten, Google, Bing, and other search engines gain extraordinary power. A marketplace's collapse does not typically occur overnight, but rather multiple phases of obstruction and illness crop up in the marketplace and slowly contribute to degradation. The marketplace of ideas may encounter disruption, based on multiple factors. The marketplace itself operates with forceful dynamics, as evidenced by new technologies, government-sanctioned censorship, journalists' failings to provide truthful and accurate information, the emergence of citizen journalism, and the growing field of "fake news."

Obstruction of the free flow of information represents an initial level of market compromise. In this atmosphere only certain goods become available for a transaction, contributing to monopoly. A few merchants rise to prominence, inculcated with power from a centralized power source. When considering information in a compromised atmosphere, a monopoly of ideas can occur. A free exchange is not permitted, precluding social, economic, and political concepts from colliding with one another to allow citizens to form new ideas. Failure represents the most intense level of marketplace compromise. In the failed marketplace, meaningful information transactions fail to occur. The market's deterioration suspends the movement of information and quells the creation of new ideas. Citizens are left wanting for the emergence of consensus on important matters, and society loses the ability to formulate collective belief systems. The absence of consensus impairs the discernment of truth because citizen consumers may fail to determine "goodness" amid the diminished choices available in the marketplace. In a healthy marketplace, power is invested into the dynamics of the market, holding the world open to various exchanges. In contrast, corruption could change the marketplace forever, shifting the power forces to an individual, government entity, or in this case a search engine operation. In the right to be forgotten environment, we see a private titan (Google), empowered and ratified by a government entity (EU), placing a choke-hold on sectors of the marketplace, strangling the free flow of information in society. This information eradication is distinguished from natural digital "decay" that occurs when servers are no longer available, hyperlinks become broken, corporations go out of business and no longer host information, and other anticipated and nonanticipated aspects of data decline and extinction occur. The de-linking presents a form of contemporary censorship and post restraint on the availability of information in a Digital Age.

KEEPING IDEAS IN THE MARKETPLACE: INFORMATION FREE FLOW IN THE GLOBAL SOCIETY

The global exchange of information occurs in real time, crossing borders and allowing citizens to learn about happenings at home and abroad. The free flow of information, long cherished by citizens of the United States, provides global audiences with opportunities to peruse the news, engage in relevant dialogues, and broaden one's individual views about a political candidate, social issue, or economic policy. The ability for readers to transcend geographic borders with knowledge and to gain information through search engines is enmeshed with the notion of free flow of information in the Digital Age. The reader has the power to procure information and "select" information of interest. Search engine algorithms, created with proprietary technology, deliver results to the individual by utilizing Big Data analytics to speedily offer up information of use and relevance to the reader. Therefore, information undergoes a distortion through the lens of a search engine. Beyond distortion, eradication occurs when information is de-linked and no longer available to the engine to recall and provide access.

While enjoying an atmosphere of readily available information, U.S. citizens often take for granted the diminished condition encountered by counterparts abroad. Government agencies in countries like China and Russia, for example, deftly censor news and information available to citizens, ripping information holes in the data tapestry and altering perceptions of reality. Consistent with this argument, the removal of information and data from global search engines is comparable to intrusive government action undertaken to censor. As we contemplate the longer-term implications of the right to be forgotten, global transparency becomes murky, allowing confusion rather than consensus to emerge for citizens seeking relevant information. The United Nations has attempted to implement general provisions about information in the Declaration of Human Rights, but independent governments operating from powerful foundations can still restrict or quell information flow across borders. Today, the global information environment is not transparent. Variable government structures can not only allow unbridled publication of information, but also limit information through censorship. A global standard for transparency of information does not exist. As we examined in Article 19 of the United Nations declaration, the text remains relevant to the current technology-driven information society and underscores the fundamental human right to have ideas shared through search engines as data is shared through other forms of media. The interference with information flow can dramatically alter the political structure and influence the power structures of democracy. In a social system that advocates the free flow of information and an informed citizenry, the audience takes an active role in questioning authority, seeking additional information, and forming relationships with others. Yet, if the information is no longer accessible, the decision-making process is imperiled. The censorship ax of the right to be forgotten imperils freedom and democracy, because media and everyday citizens no longer have access to an accurate recollection of facts and occurrences in public life.

The Law and Marketplace Preservation

While the European Union implements legal actions that have an impact on the free flow of information, the United States First Amendment squarely opposes government interference through prior restraint on revealing information or executing censorship on specific information. However, the advent of fake news, unauthenticated journalists, fabricated media outlets, and a laissez-faire approach to the publication of libelous materials has created a convoluted marketplace. Some scholars argue that some sort of avenue for digital redemption could answer the calls for de-listing in a palatable way (Ambrose, 2016), yet others assert these movements quell free flow of information and violate generally accepted communications ethics codes (Byrum, 2017). A global standard is required for regulation of content in the right to be forgotten atmosphere. The EU contends it has taken actions to protect citizens and give them a semblance of control over their personal information online. In a cursory examination, such actions may seem better fit for considering erroneous or outdated information. Further, citizens seeking de-linking from revenge pornography sites may gain a reprieve in the European Union. Yet, in the United States, adherence to the First Amendment portends a different atmosphere, one that takes a laissez-faire attitude to the marketplace of ideas with a broad acceptance of information published and linked on various websites and retrievable via popular search engines. While conflict between privacy and transparency leads to robust debate, we now face an imperative to more effectively manage the global marketplace of ideas. In a right to be forgotten environment, how can authenticity and transparent communication endure? Further, with the absence of a global standard of information regulation and compliance, citizens of various countries must choose from a buffet of different information. The information retrieved by a citizen in the United States is altogether different than the retrieval results in Europe. For Europeans who undergo the right to be forgotten procedure in the Europe and successfully de-link information may experience a false sense of security, as a United States friend (or enemy) can easily search for the information and, if linked via a United States–based search engine domain, access the information.

The Marketplace of the Future

By returning to the analogy of the marketplace of fruits and vegetables, we may comprehend how the principles of free market economics come into play in the twenty-first-century marketplace. There is no question government intervention to snatch a basket of apples from the inventory means they are erased from inventory, and shoppers must access other alternatives—or have nothing at all. Let's apply this apple idea to a concept on the World Wide Web. Consider a story presented online about a particular individual linked to a search engine. The individual can petition the engine for dynamic and efficient removal, forever changing the results available, as if it never existed. While European audiences may applaud this ability as a victory for individual rights and the empowerment of the citizen to have control over her or his own image, there are far-reaching implications for society by eliminating online data, particularly when considering the First Amendment and freedom of speech. The specter of government intervention and intrusion into the lives of American citizens remains in check because of the United States Constitution, yet the right to be forgotten invites government into personal data, allowing for a paternalistic, governmental view to decide what is "right" to be presented online. For advocates of the First Amendment and champion of the free flow of information, the right to be forgotten foils the marketplace of ideas, because it creates memory holes, articulates incomplete information, and obscures transparency, leading to far-reaching implications for both citizens and journalists who attempt to accurately report societal happenings. Further, in a representative democracy, the process of vetting a political candidate by evaluating the individual's past represents a vital task. The right to be forgotten impinges on this act of comprehensive, accurate evaluation and instead allows for revisionist history, complete with memory holes. The next chapter addresses various dimensions of the role of communication in democracy and shows how the right to be forgotten compromises not only the evaluation process for specific individuals but also influences the dialogue about public issues. No matter what, just as the fruit stand must stay open to feed the villagers, the marketplace of ideas must remain open to allow for transparent political, economic, and social structures, both in the United States and abroad.

Chapter 3

Communication in Democracy

When an American voter steps into the voting booth, the individual brings with them a specific body of knowledge, a cadre of beliefs developed through a lifetime of experiences and perceptions of the world around them, changed by exposure to various information sources. The cacophony of media sources churns much louder and faster than the penny press of Colonial America, yet the adherence to the rich heritage that binds freedom of speech and freedom of the media to democracy remains. The individual voter today may temporarily silence a cell phone to push the digital levers on the ballot, but the open world of information courses through the digital arteries of the person's smartphone, tablet, or computer. Today, active voters can passively receive information and pursue information about political candidates and relevant political issues both as a conscious receiver of political information. At other times, the information comes over the transom in a way that does not allow for conscientious processing, but more appropriately could be classified as passive information. No matter how the information gets to the American voter, the presumption remains that the information is available to procure. The five rights itemized in the First Amendment (freedom of speech, press, assembly, association, and petition) have been called the linchpins of American democracy (Bhagwat, 2016), thus deserving of protection from government interference. This chapter indicates how the European right to be forgotten contravenes the basic freedoms found in American democracy. While American politicians and political candidates still can suffer the eye of scrutiny found in a right to be forgotten-free environment, the larger, global informed citizenry may face a new predicament. The de-linking of information chisels away at the totality of available information, leaving voters in a position that could be manipulated by a power source, in this case a private data controller. We live in a world of porous boundaries between political and social culture, converging on digitally connected devices like a smartphone. Citizens have become reliant on the information available at their fingertips.

Formation of public opinion occurs through the transmission and reception of information. Public opinion encounters a complicated, confusing journey when information from multiple authors travels through the Internet to search engines and ultimately arrives in the eyes and ears of the recipient. The ability of public opinion to gain a critical mass differs from the heritage of coffee shops devoted to various trade groups and political parties. In fact, some would argue that the World Wide Web is analogous to a gigantic coffee shop, shopping mall, or robust marketplace that allows for the exchange of ideas and formulation of public opinion. The notion of public opinion appeared in the seventeenth century with John Locke's writing, *An Essay Concerning Human Understanding* (Locke, 2001). U.S. Constitutional father James Madison recognized the imperative for an educated citizenry who possessed the capacity to form collective opinions and envisioned an enlightened majority would have the capacity to mobilize against an illegitimate regime (Gabrielson, 2009). By deconstructing public opinion, we can observe the impact of political opinion on society. For example, beliefs comprise the basic building block of public opinion, and, if these beliefs influence our behavior patterns, they become an attitude. These attitudes graduate to an opinion when we share them with others, and public opinion arises when a multitude of people share the same opinion (Guth & Marsh, 2012). As a result of this sequence, a reader who retrieves information on a search engine begins to develop beliefs and attitudes from the immediate rendering of the search engine results. If the rendered information is riddled with holes, the reader's ability to form a public opinion is crippled.

THE PUBLIC SPHERE AND PUBLIC OPINION

Philosopher Jurgen Habermas pointed to the public sphere as a place for discussion of public events. In his seminal book, *The Structural Transformation of the Public Sphere—An Inquiry into a Category of Bourgeois Society* (1991), Habermas asserts the public sphere acquires an identity as a place where individuals can unite, discuss collective social problems, and, through that discussion, ultimately influence political action. Reminiscent of the Greek city-state (polis), the public sphere affords the opportunity for public debate and formation of opinions. Habermas's work, recognized as a foundation for subsequent theories on the public sphere, characterized the decline of the political sphere and the rise of the changing forces of media. These crucial forces included the introduction of commercialism and private gain into the public sphere, transforming the bourgeois public sphere and adapting the way that citizens address political concepts. As Habermas documented the transformation of the public sphere to greater commercializing, a new emerging field grew. The public relations profession entered the scene as a leading force to influence and shape public opinion. Edward Bernays, who is often characterized as the father of public relations and who was the nephew of Sigmund Freud, provided the world with the book, *Crystallizing Public Opinion* in 1923. Bernays coined the term public relations counsel, providing a viable alternative to negative connotations including "propagandist" or "publicist," and through his writing the process of two-way communications matured in practice (Bernays, 1923). The two-way communication paradigm provided a solid foundation for subsequent communication models, including those prevalent in contemporary public relations theory and practice (Grunig & Grunig, 1992).

In historical scholarship about the formation of public opinion, noted luminaries like Bernays and Walter Lippman, the author of *Public Opinion* (1922), arise. Writing during the same historical period in the 1920s, Lippman is credited with a critique of public opinion and the role of citizens in forming opinions in a democracy, while Bernays underscored the significant influence of thought leaders in society, those who help shape and determine salient discussion points for society. In their early writings, neither Lippman nor Bernays may have anticipated the technology-enabled atmosphere that pervades modernity, yet both contribute theoretical insights that resonate today, as we contemplate the role of citizens and communication in society. Lippman writes, "The world that we have to deal with politically is out of reach, out of sight, out of mind. It has to be explored, reported, and imagined. Gradually he makes for himself a trustworthy picture inside of his head of the world beyond his reach" (Lippman, 1922, p. 20). The casual smartphone user who receives tweets or brief alerts from news organizations and occasionally accesses a search engine to find more in-depth information about a specific topic is using these technology tools to create what Lippman called the "picture inside our heads." In his critique of mass media, Lippman suggested media are susceptible to manipulation and they can signal the occurrence of an event through the journalistic process of the selection of news and the discernment of societal happenings.

As we contemplate the formation of public opinion, Bernays's seminal writings suggest a role for an opinion leader to manage the movements of the "herd," as they gather public opinion. Bernays suggests the passivity of viewers results in the formulation of public opinion by simple exposure to mass media outlets. In a chapter of *Crystallizing Public Opinion* (1923), Bernays addresses the role of media of the time, including newspapers, posters and advertising: "People accept facts which come to them through existing channels. They like to hear new things in accustomed ways. They have neither the time nor the inclination to search for facts that are not readily available to them" (Bernays, p. 146). Consistent with this tenet, we recognize that today's social media has become "mainstream" and provides information that is accurate sometimes and inaccurate sometimes. Citizens today gain information as it actively pours over the transom of the smartphone, buzzing, beeping, and bearing characters with tiny bytes of information. Further, Bernays suggests that readers operate passively, letting their personal opinions gain shape by what enters their lives. This situation seems relevant considering the inclination of Digital Age citizens to gain no more information than what arrives in front of their eyeballs from a media channel. This could be a television station, newspaper, or online media, or morsels of information from another individual in the form of a tweet, a text or a shared hyperlink to a news story, often accompanied with a pithy statement of opinion. In contemplation of the right to be forgotten, the "average" citizen may not care to use the search engine in a meaningful way, so why is it imperative to retain information on the search engine at all? In the same vein it may be argued that citizens of the world use search engines about six billion times a day to gather information and curate their own "world" of information.

While Digital Age citizens may conduct an ambitious search for information by tapping search terms into Google and the like, information continually streams into their consciousness through email, social media, mobile applications, and other message carriers in the digital milieu. This creates a modern-day manifestation of a condition that Jurgen Habermas diagnosed, indicating the decline in the public sphere and the naiveté of citizens to simply embrace the "publicity" and stunts of communicators in multiple realms, most notably the political sphere. A stylistic show has emerged in Digital Age democracies reflecting the richness of the media production channels. Individuals project a face on social media; some provide an authentic reflection while others distort. Habermas has written extensively on the decline of the bourgeois social sphere as a venue for a spirited public debate that allows for the formation of public opinion and open dialogues among citizens; he points a chastising finger at public relations practitioners and "advertisers" as culprits in the demise. Specifically, Habermas attacks staged publicity as a "transmutation" that pervades the entire public sphere in the political realm:

> Before the expanded public sphere, the transactions themselves are stylized into a show. Publicity loses its critical function in favor of a staged display; even arguments are transmuted into symbols to which again one cannot respond by arguing but only by identifying with them. (Habermas, 1991, p. 206)

Habermas further points to the employment of advertising in the mass media-dominated public sphere as a way to influence votes and, in the face of this political marketing, the electorate disintegrates as a public (1989, p. 217). Does the online sphere remains analogous to a "world" where individuals seek information, exercise access, formulate public opinion, and engage in the physical world through social, economic and political conditions? Or rather, does the passive viewer simply absorb ideas as they come over the transom, failing to exert rational questioning or question the fluid assumptions of arguments? Research suggests distrust is on the rise in the atmosphere of populism, characterized by a media "echo chamber" that allows for the reinforcement of existing attitudes and beliefs.

The Edelman Trust Barometer, performed each year measures global trust of four institutions worldwide, including business, government, NGOs, and the media, survey individuals in 28 countries. In 2017, the study of 33,000 respondents found that, in their eyes, trust is declining across the board, with media suffering a precipitous decline from the previous year (*2017 Edelman Trust Barometer*, 2017a). The majority of the study's participants (53 percent) indicated rising distrust in light of a common belief that institutions are failing them. Edelman's study reveals insights into growing populism and the decline of news media as an institution that merits trust. "The implications of the global trust crisis are deep and widening…the consequence is virulent populism and nationalism as the mass population has taken control away from the elites," Richard Edelman, president and CEO of Edelman, stated in a news release (Edelman, 2017b). Edelman suggests lack of faith in the system, combined with fears in the United States, the United Kingdom and Italy, led to the election of Donald Trump, the Brexit vote, and the failed Italian referendum (Edelman, 2017b). The 2018 Edelman Trust barometer continues to document the decline, reporting that trust in the U.S. suffered the largest-ever-recorded drop, as faith in government collapsed (Edelman, 2018). One year later in a prepared statement, Edelman stated, "The United States is enduring an unprecedented crisis of trust."

LIVING IN AN ECHO CHAMBER

As a consequence of this declining faith in the system, citizens are handicapped in receiving accurate information, based on the survey. In what Edelman has deemed the "echo chamber," citizens now seek information that reinforces personal beliefs while they eschew opposing points of view from reaching their awareness. The study found that respondents favor search engines over human editors and likely ignore information that disagrees with their belief system (Edelman, 2017b). This finding is particularly significant in consideration of the European right to be forgotten, because the practice endorses the data removal from search engines, even though these tools generate more trust than traditional media or social media. The Edelman report indicates 64 percent trust in search engines, compared with 57 percent for traditional media and 41 percent for social media (Edelman Executive Summary, 2017a). This failing global trust implicates a changed dynamic for the news media. "People now view the media as part of the elite. The result is a proclivity for self-referential media and reliance on peers. The lack of trust in media has also given rise to the fake news phenomenon and politicians speaking directly to the masses," said Edelman in a prepared news release (Edelman news release, 2017b). Edelman uncovered a "further dispersion of authority," indicating that average citizens are now just as credible as sources of information, when compared to technical and academic experts. This study pulls back the curtain on the growth of citizen journalism, direct communication, and the deterioration of traditional journalistic theories that upheld values of objectivity, ethical adherence to the truth, and acceptance of news as worthwhile, because it would successfully travel through a series of decision-making gates, transforming the information to a state of "news." Edelman's chief content strategist, Steve Rubel, offers this point of view: "The emergence of the echo chamber is directly correlated to the dawn of the Age of Technology platforms. Now that these platforms are the primary way we discover and consume information, it has made it easy for billions of people to tune more deeply into proximate peers, tune out all others, and validate their worldview" (Edelman, Executive Summary, 2017a). We live in a world exposed to "fake news" that distorts and reinforces individual and collective judgment, a true danger to rational thought and its political fruit, democracy. The advent of social media has fueled a new way of "knowing," one that does not rely on the accurate reporting of facts.

Disruption of media channels by technology does not represent a new phenomenon. The First Amendment's constitutional protections have endured the advent of mass printing presses, the formulation of radio, the creation of television, the emergence of network television, the genesis of the Internet, and the fragmentation of content into blogs, video channels, websites, mobile content, and now social media. First Amendment scholars have traditionally underscored the heritage, practice, and inexorable link between the free flow of information in society and a working, representative democracy. Theorist Alexander Meiklejohn, founder of the Meiklejohnian Theory of the First Amendment, resoundingly underscores the justification for free speech, because of its valuable role in informing citizens who would in turn exercise their democratic society through voting. "First of all, the freedom to vote, the official expression of a self-governing man's judgment on issues of public policy [must] be absolutely protected," Meiklejohn writes (1961, p. 256). Yet, Meiklejohn pushed further in the protection by stating:

> There are many forms of thought and expression within the range of human communications from which the voter derives the knowledge, intelligence, sensitivity to human values: the capacity for sane and objective judgment which, so far as possible, a ballot should express. These too must suffer no abridgment of their freedom. (Meiklejohn, 1961, p. 254)

The rights itemized by Meiklejohn specifically encompassed education, the achievements of philosophy and the sciences, literature, arts, and public discussion of societal issues. Further, Meiklejohn affirms the audience's role to choose relevant information. When responding to calls for censorship, Meiklejohn writes: "Here as elsewhere, the authority of citizens to decide what they shall write, and more fundamentally, what they shall read and see has not been delegated to any of the subordinate branches of government. It is 'reserved to the people,' each deciding for himself to whom he will listen, whom he will read, what portrayal of the human scene he finds worthy of his attention" (Meiklejohn, 1961, p. 262). This opinion pays homage to the marketplace of ideas theory of the First Amendment and casts light on the current predicament catalyzed by the European General Data Protection Regulation. The European right to be forgotten empowers Google, Bing, and other search engines to function as data controllers, inculcating a specific individual into a powerful position to vote "yea" or "nay" to the inclusion of information. This bold move in Europe counters the spirit of the First Amendment in the United States. In a Meiklejohnian world the American citizenry, as the "people," exercise a right to gain access to information and participate in contemporary dialogue about a multitude of topics to help them gain knowledge to equip themselves as informed voters in a representative democracy. Further, some scholars have suggested Meiklejohnian Theory places expression that relates to the self-governing process in a different class for protected freedoms, providing ultimate First Amendment protection (Abrams, 2017). Considering this phenomenon, the First Amendment becomes the North Star to finding a way to preserve the will of the people and democracy.

The ability to forecast what data will be considered as newsworthy information in a future time is impossible. The hypothetical crystal ball of discerning what is no longer "relevant and outdated" cracks when we consider that something that is not newsworthy now may in the future become a significant newsworthy event. When it comes to the right to be forgotten, a data controller is now charged with effectively discerning what information is relevant to a voting decision. One individual who works for a search engine in the right to be forgotten help desk does not constitute a public that operates in the representative democracy. We must exercise caution. A data controller should not make a willy-nilly decision about an individual at any given point in time that may accrue greater cultural consequences at a later date, if that individual enters public service or thrusts themselves into the limelight to support a particular cultural issue. While the tenets of the Meiklejohnian Theory resoundingly affirm free speech because of its role in self-government, contemporary scholars suggest new ways of contemplating the First Amendment and freedom of speech in America, particularly in light of the Internet as a communication channel, search engines as data controllers, and the archiving of information in the form of Big Data. The notion of the vital link between the free flow of information and self-government has been defended by multiple First Amendment scholars (Bork, 1971; Post, 2011; Sunstein, 1993; Weinstein, 2011). Questions still emerge about the effectiveness of the constitutional right for all citizens, when we consider access to the media channels and the Internet; the government's ability to withhold information from the citizens; and private ownership of news media and distribution outlets (Sunstein 1993).

ADVANCING DEMOCRATIC INFORMATION THEORIES INTO THE DIGITAL AGE

With the legacy of the democratic theories of the First Amendment weighing in on the comprehension of the role of free speech in society, scholars continue to advance understanding of the part free speech plays in democracy, while the world's citizens remain overwhelmed by an atmosphere of constant data retrieved through search engines, pushed via social media giving individuals a saturating daily dose of information. These circumstances have caused constitutional theorists to reframe the First Amendment, as Bhagwat (2016) indicated in characterizing the First Amendment as "kaleidoscopic." This advances the concept from the narrow democratic view that freedom of information should have protection because of its role in informing the electorate, to a more inclusive position that freedom of information must be guarded because of the multifaceted role citizens are destined to play in a democracy: challenging elected representatives, communicating values, and forming alliances through assemblies (Bhagwat, 2016).

The result of digital culture has transformed the theoretical landscape for freedom of information, as indicated by the emerging theories and new interpretations of the First Amendment. Yet, the Digital Age calls for a more inclusive model that encompasses popular culture and considers various facets of democratic culture. Responding to this call, Jack Balkin has set forth a model of democratic culture that encompasses both the democratic elements and corresponding cultural elements (2016). Balkin outlines the inferiority of democracy-based theories in the realm of digital speech and digital networks and cites three reasons: the constitutional protection of the art and the freedom to engage in artistic expression in digital media; the structure of global discourse, which transcends the nation-state; and the global nature of the Internet, which can be undermined or threatened by national regulation of telecommunications media (Balkin, 2016, p. 1089). His theory about democratic culture asserts the World Wide Web gives rise to new spheres of public influence that may likely transform into new global institutions and practices. While warning against the prognostication of a unitary global sphere, Balkin states that if we define public discourse as a circulation of ideas and opinions, then it is already well under way. This global public discourse operates on the assumption of communications architecture that provides a fertile environment for the sharing of information both inside a nation-state and beyond. Balkin asserts:

> In democracy-based theories, one has to show that a nation's Internet regulation harms self-government within that nation-state. In a theory of cultural democracy, by contrast, one need only demonstrate that Internet regulation harms the global exchange of information, art, opinion, and ideas. (Balkin, 2016, p. 1094)

Consistent with this thinking, we can extrapolate a key lesson about infringement on the cultural democracy theory. The right to be forgotten, a form of online regulation, harms the global exchange of ideas and thus counters the cultural democracy theory.

Working to promote the free flow of information outside the United States may assist an informed public inside the country. It might help give people within the United States a warranted belief that American officials will be responsive to their concerns (Balkin, 2016, p. 1092). Balkin points to the right to be forgotten as a threat to the global public good as evidenced by the fact that the European Union now makes it more difficult to share information across the world (Balkin, 2016). The advent of a digital democracy remains nascent, based on the fact that while citizen blogs have grown each day, the ability to capture eyeballs and persuade individuals remains in the balance. To better understand the rising political power, observers must examine access to online information and statistics of blog readership (Hindman, 2009). The European right to be forgotten stands counter to the democratic self-governance streams of thought of the First Amendment, both historic and contemporary. The right to be forgotten by its function of creating memory holes, distorting history, and adapting thinking offers a chilling effect on global culture.

Watchdog Role of Media

While the information ecosystem in the United States operates on the premise of the free flow of information in society, the call for transparency and accessibility of information comes from strong ideological roots in the nation's founding and practice of journalism in the country. The government watchdog, the fourth estate with its power of the "pen," is commonly used to describe the role of a mindful media holding each elected official accountable for public (and sometimes private) actions. On an institutional level, the healthy functioning of the fourth estate waves to the inclination that the media may function as a "fourth branch" of government, providing the checks and balances required for a healthy democracy. The watchdog role gained respect for news media, which often advocated for societal interest and gave power to disenfranchised citizens, documenting the real story behind society's dark side and operating with a dogged pursuit of the truth. A healthy media performs an important societal function, solidified by objective news reporting, transparency, and fairness. The Society of Professional Journalists, founded in 1909, publishes an ethics code that reflects this same tonality. Young journalists may enter the profession with the dogged passion to uncover misdealing and accurately report the truth. The watchdog function has ebbed and flowed through time. Watchdogs' publicly perceived "victories" may include journalistic exposés by the *Washington Post* that precipitated the resignation of Richard Nixon, and the *Boston Globe's* investigative series that exposed sexual abuse by priests in the Catholic Church. The media conduct various activities including fact-checking, beat reporting, and investigative journalism to shine a scrutinizing light onto public officials open to public inquiry, political happenings, and societal concerns. In a global study of journalists, researchers have found the emergence of the "detached watchdog" as one of four contemporary journalism cultures. Journalists who subscribe to the detached watchdog role function with its overarching tenets of neutrality and objectivity, casting aside tendencies to become involved or intervene in a situation. The ability for the reporter to actively witness happenings and report on them reflects a long-standing theoretical correlation with positivism and corresponding alignment with objectivity (Schudson, 1990).

Without objectivity, what principles guide the contemporary practice of journalism? The proliferation of citizen journalism, the rapidity of the 24/7 news cycle, the failure of some practicing journalists to verify basic facts, and the ubiquity of direct communication via social media warp the contemporary environment for objective journalism. The role of the "watchdog" has also adapted in the fragmented Digital Age. The eyes of the citizen now focus directly on the politician, with individual citizens engaged in public life. The "monitorial habit" operates with the tool of the online search engine, making the right to be forgotten a threat to the integrity and clarity of the contemporary information gathering and reporting environment. Yet, this watchdog function is broken. According to a recent study, six in 10 people believe search engines more than human editors (Edelman Trust Factor executive summary, 2017a). Citizens now look to social media for information, exposing themselves to data that is not checked for accuracy or vetted by a legitimate news source.

Search Engine Data Controllers as Contemporary Gatekeepers

This media watchdog concept has proliferated amid an environment in which the news media operated with objectivity to report the facts. Watchdog journalism matured in tandem with communication theories like gatekeeping and agenda-setting. We may find relevance to the current right to be forgotten situation by looking back at the theories. The Gatekeeping Theory offers the premise that messages are created from information about events that has passed through a series of gates and has thus undergone change as a result of the process (Shoemaker & Vos, 2009). Media scholars traditionally asserted gatekeeping determines a person's social reality and view of the world (Shoemaker & Vos, 2009). Consider editors on city desks throughout the United States who once determined the news of the day for their community and chose to allow certain information through the "gate"—or alternatively, to turn the information away at the gate and quash a story. Editors, producers, and reporters—everyone on the journalistic food chain—made critical decisions about what passed through the gate. To achieve the role as a gatekeeper, the individual often underwent specialized education in one of America's journalism schools, joined a professional organization like the American Society of Newspaper Editors, the Society of Professional Journalists, or Women in Communications, and adhered to a basic modicum of journalistic ethics grounded in both education and practice. However, today the gate stands wide open on the Internet. Journalistic gatekeepers no longer stand guard on the dissemination of "truth," but rather untrained "citizen" journalists can now publish data immediately through social media distribution. Criticism of the Gatekeeping Theory points to the fact that events not covered by news media often remained inaccessible to potential audiences. The prevalence of citizen journalism and firsthand accounts pushes open the door to a new level of mediated experiences for social and political purposes. This dissolution of the filter has led to an explosion of data into the marketplace of ideas, but some of the information may suffer from falsehoods or mischaracterizations. While we consider the notion of gatekeeping, the appearance of the search engine as data controller gives rise to new implications, characterizing Google, Bing, and other search engine operators as the omnipotent Digital Age gatekeepers to determine the availability of public information.

We may also find the gatekeeping function resides not only with the search engine, but also with the individual media consumer. To understand this better, we can look back to the heritage of Gatekeeping Theory, developed by Kurt Lewin, a pioneer in applied psychology in the United States. Lewin looked to early twentieth century thinking from Gestalt psychology and Husserlian phenomenology to Freudian psychology when developing the gatekeeping concept, honing on factors that influenced the individual's thought processes, environmental perceptions, and unconscious decision-making (DeIuliis, 2015). Thus, contemporary literature suggests Gatekeeping Theory emerged from conceptualization whereby the social world was a relationship between individuals and groups. Communication scholar David Manning White applied Lewin's concepts to the newspaper business when he analyzed "Mr. Gates" and his treatment of stories from various wire services, leading White to conclude these decisions were highly subjective (White, 1950). Today the traditional gatekeeping channel, embodied by Mr. Gates, is now performed by the individual, working in concert with a search engine, social media, or other technology-aided device to discern news. A recent study of 2,000 adults in the United States using online sources for news found that people were most likely to obtain news from a news organization website or app, social media, or search engine (Pew, 2017). Further, respondents were likely to remember the source of the news when they clicked on a hyperlink to gain more information; however, they were less inclined to remember the source, if it came from social media or a friend (Pew, 2017). These findings suggest individuals still assign value to the reporting of news organizations, yet the source becomes convoluted when it arrives through social media.

Historically, distinctions have been made between newsgatherers and news processors, differentiating between reporters and editors, journalists and producers (Bass, 1969). Today, the uninhibited realm of content creation by citizen journalists allows for the exponential growth of newsgatherers in the information communication process and diminishes the activities of the news processor, because information streams directly through the Internet medium to audiences without a filter or monitoring. Under a right to be forgotten environment, individuals may argue that the search engine becomes a filter of last resort or remedy in the case of information no longer relevant—yet this filter is not a journalistic one and not a news processor, but rather a subjective data controller. The subjective decision-maker may have a policy and guidelines, but the critical choice to remove information is relegated to an individual working for the search engine. These digital concepts of information control pave the way for a discussion about the evolution of the societal "public" and the corresponding formation of public opinion under these circumstances. Data controllers become digital gatekeepers that impinge the public's right to gather information, share opinions, and virally spread information, and this can be as chilling as police in a public square regulating or interfering with public assembly. Further, the articulation of new ideas from this process could lead to adaptations in the political sphere by changing the way citizens receive and process information.

The notions of objective journalism have given way to a new vernacular popularly articulated as "transparency." The term transparency pervades dialogues of all types related to economic, social, and political happenings. Citizens clamor for transparency to shed light into these situations. This text argues the right to be forgotten makes transparency opaque because the practice warps the ability to see through a situation because it is manifest with holes. The following section addresses how the right to be forgotten contravenes the First Amendment and the corresponding theoretical frameworks of the role of freedom of information in a representative democracy.

RIGHT TO BE FORGOTTEN INHIBITING FREEDOM OF INFORMATION IN DEMOCRACY

In a democratic culture embracing the journalistic function as one that serves a watchdog on government, the right to be forgotten handicaps print, broadcast, and online journalists who attempt to analyze contemporary circumstances and document happenings in society. Reporters turning to Google, Bing, and other search engines to access information from the European domains must recognize the results may only tell a portion of the story, leaving details about an individual in the abyss of oblivion created by the right to be forgotten. So we may question, why is the absence of an accurate portrayal of the individual so important moving through time? In consideration of the arrested Navy Captain with a wrestling fetish, one may wonder whether or not this would be of concern to a voter if the captain attempts to enter political life later in his career. When the item remains properly de-linked, it may not readily appear again. In contrast, an industrious reporter may unearth the information using traditional investigative techniques, so that proper evaluation of the candidate can occur. In the United States, the journalistic vetting of candidates often uncovers minute facets of an individual's personal and professional capabilities, past transgressions, and current wrongdoings, allowing the citizenry to hold them accountable and make an informed decision when casting a ballot. This accountability resides on a spectrum that can change from season to season or vary among political parties. The issue should not be whether the information is comprehensive and may be "too personal" or irrelevant—the salient point is that the information *is available*. Some may argue information too personal or that does not reflect the person's fitness to serve should be fair game for deletion. However, in a democracy, information published before public eyes can be viewed by all citizens, giving voters the opportunity to make their own cogent decisions about the candidate and to exercise an individual right to endorse or deny the candidate. While de-linking "removes" the information from a specific search engine, the data may still reside in the news organization's website and remain catalogued in another way. A second threat to the search engine de-linking in society is the phenomena of obscuring an issue or altering the perception of an event. A third threat to search engine de-linking is the interference with the interpretation of societal trends through absence of data or clarity.

The distortion of events occurs not only in the manifestation of the event itself, but also in the altering of the collective memory of the event. The reach of a protest often transcends those individuals who attended the demonstration to a far-reaching audience composed of participants who tap into the event through search engine recovery of news stories and other coverage of the event that may have been generated by consumers or other citizens. This expands the scope of the protest's impact, which represents political activism in the Digital Age. The right to be forgotten has the opportunity to change the perception of the event. Let's consider the example of a controversial student's participation in a protest advocating for gun regulation. The student is arrested for trespassing during the protest and after several years seeks to remove the hyperlink to the arrest report listed on the search engine. If Google honors the request, the student may now be able to successfully secure a job unencumbered by the fallout from the news story, but the historical interpretation and evaluation of the event experiences change. Citizens researching this event can become ill-informed about the story of student arrests, because the information is not immediately available. Moreover, once again, a gap occurs in the individual's history. If this person attempts to run for office in the future, various constituencies would eagerly embrace information about an arrest at a past protest about gun regulation. Thus, we can conclude that the right to be forgotten obscures the collective perception of cultural events, most notably the practice, participation, and memory of protest events. The suppression of memory of these protest events warps and alters the cultural understanding and thus perceptions of particular social, economic, and political issues. By changing the individual's public persona available online, resulting effects occur. Some may argue the ability to remove this data denotes a privacy right that honors an individual's right to manage available online data, while this book asserts the removal interferes with the free flow of information in society. The free flow of information in society not only faces threats from removing content by an individual, but also threatens the collective cultural and historical memory of specific events, including protests and political actions. Lastly, the removals at the hands of the powerful search engines impair the ability of journalists, sociologists, cultural anthropologists, and others performing research in the social sciences to effectively observe societal trends moving through time. In sum, the right to be forgotten's obstruction of the uninhibited free flow of information leads to consequences beyond the individual level to influence society at large.

The full range of journalists, sociologists, psychologists, and others interested in society's shared experiences stand to lose a valuable resource for information in the right to be forgotten atmosphere. Specifically, the documentation of societal happenings becomes flawed, based on the interference of a data controller required to comply with the Data Protection Regulation. To illustrate this point, let's consider the removal of data linked to various crime statistics. For example, over a five-year period, a local village experiences 15 sexual assaults, thus revealing an incidence rate of three assaults per annum. A reporter can gain access to the original news stories about the events or tap into aggregate information released from the police department. In the event of search engine de-linking, the rich details of the original story and the resolution of individual cases suffer "memory holes." The perforated memory of these events prevents reporters, medical professionals, potential victims, and others who have a stake in analyzing societal trends from understanding the totality of circumstances around the events and the cultural impact.

Under the right to be forgotten, private enterprise (Google, Bing, and other data controllers) assumes a powerful role in adapting perception. For American scholars, this interference in information by a public source conjures questions related to the Freedom of Information Act and other legislation designed to protect the integrity of the free flow of information in society. In addition to the First Amendment, the United States has fortified a position of free flow of information and government transparency by the creation of subsequent laws. Consistent with the compelling need to maintain transparency, the U.S. government enacted the Freedom of Information Act in 1996. Commonly called FOIA, the law grants citizens and journalists access to information held by the federal government and its corresponding agencies. While the legislation provides a rationalization for reporters and the public to gain access to information, the act itself covers nine exemptions, including national security; internal agency rules and procedures; disclosures forbidden by other statutes; trade secrets; agency memoranda; personal privacy; law enforcement records; financial records; and geological information (FOI.gov). The law subsequently expanded in 1996 with the supplement of electronic Freedom of Information that covers electronically stored information. In addition to records, individuals have access to government meetings. The Office of Government Information Services (OGIS), part of the National Archives and Records Administration, serves as a mediator for the FOIA conflicts that may arise when an agency denies a petition for the information.

To grant the public and media access to meetings of the federal government, the Government in Sunshine Act was enacted in 1976. Also known as the Federal Open Meetings law, this law requires agencies to provide advance notification of meetings and document happenings through minutes (5 U.S.C. 552b). Many of these laws have been duplicated by various states.

These laws underscore the fundamental value to protect the free flow of information in a democracy by documenting the activities of individuals and federal agencies acting on behalf of the citizens. United States citizens value the notion of transparency, as evidenced by these laws designed to hold the government accountable. In a right to be forgotten atmosphere, even information that proximally interferes with these protections of freedom of information would certainly meet a wall of opposition as threat to transparency. While the Freedom of Information Act empowers citizens, journalists, and other information seekers to gain access to federal documentation, the same type of protections do not impact private enterprise. In fact, corporations do not have the same disclosure requirement as federal agencies operating with taxpayer dollars; however, as a publicly traded company, Google must file quarterly and annual reports, adhering to the Securities and Exchange disclosure standards about the financial performance of the company. From the democratic citizen's ability to gain information, the United States derives a corollary concept—the Right to Know. This right has given the citizenry an unabashed fervor for the availability of information and an intense spirit to "get to the bottom of things" through investigation and information. The following section addresses the citizen's role in a democratic society.

The Citizen's Role

As we consider the role of the citizen in an active democracy, two relevant themes merge. The citizen embraces a duty as an active information-gatherer and now in a newly emerged role, the citizen becomes the message carrier, often performing the role of a citizen journalist or "thought leader" who sculpts public opinion by weighing in on various topics and sharing viewpoints via social media. Both of these roles put the citizen in a dynamic situation that presumes the individual has access to the required information to viably gather data and form beliefs worthy of sharing. This prevailing need for information may also earn the earmark as "the public's right to know," an adage of American journalism and politics to underline the society's need for transparency among various organizations in the political sphere and beyond. As a result, as active listeners in the happenings of the digital world, information consumers assume they have a "right to know" every granular detail of information that has ever appeared on a major search engine. Once again, though it is noted that privacy activists would assert that privacy rights should prevail in some instances over the public's right to know. Internet users predominantly embrace the notion that once information has been posted, it can appear repeatedly along multiple social media channels. The failure of data to render would indicate that censorship occurred or a natural form of digital decay took place. Citizens are left wondering, "Where does the public's right to know end and the individual's right to privacy begin?" This topic is addressed in more detail in chapter 6; however, the right to be forgotten argument often falters in the face of First Amendment advocacy supporting the concept of an informed electorate.

The ubiquity of the Internet and the global spread of social media channels have led to the creation of a society of self-disclosure (Thompson, 2005). This communication-intensive world untethers individuals from a common location, and we must no longer be present physically to witness a happening. Some politicians and activists successfully leverage this mediated presence via public relations consultants and communication advisers to broaden support, awareness, and virtual presence among multiple constituencies. The role for the citizen has also adapted to the Digital Age. The web-enabled scope of transparency contributes to the monitoring of democratic events. The emergence of the multimedia society has led to the formation of the monitorial democracy, enhancing the level of scrutiny on public bodies (Keane, 1991; Schudson, 1998). As monitorial citizens, the Digital Age netizens use commonly available search engines to locate information and form opinion.

The second stream of the argument addresses the complex notion of the citizen as a content creator. Professional trained journalists were once the primary gatekeepers who provided the public with vetted news stories presented in an environment with trust and accuracy. The gate is now wide open with citizen journalists that proliferate online venues with "news" information. According to Technopedia, "citizen journalism refers to the reporting of news events by members of the public using the Internet to spread the information. Citizen journalism can be a simple reporting of facts and news that is largely ignored by large media companies" (*Technopedia*, 2017). The Pew Research Center reported that crowdfunding sites like Kickstarter have become popular funding resources for startup journalism enterprises, including magazines, books, and documentaries (Vogt & Mitchell, 2016). The phenomenon that every single person with a smartphone, computer, or web-enabled device creates content susceptible to World Wide Web archiving and search engine registration demonstrates the ubiquity of the digital information marketplace. Citizen content creators proliferate ideas in the marketplace, including some that individuals in Europe would seek to have de-listed in a right to be forgotten claim. In the United States, adherents to freedom of speech tenets assert that citizen content deserves protection, unless it is presented in a way legally classified as libel. In that case, the plaintiff could seek a traditional remedy of relief by proving the tort of libel, thus demonstrating publication, identification, falsity and fault (Lewis, 1991). In the United States, Google sets a high threshold for removal of content and will not do so without a court order finding the statements are defamatory (Donovan, 2017). We observe a new communication landscape today, where everyone is a content producer. In addition to the active role of the citizen as a consumer content creator and thought leader, individuals function today as news aggregators, digesting news and information and then distributing the data to a universe of followers in the social media sphere or through more traditional online channels. According to the Google Transparency Report (2018) between 2014 and March 2018, the company had received 655,561 requests to delist with requests for de-linking at 2.4 million (Google, Transparency Report, 2018). According to the report, France's annuaire.118712.fr and www. Facebook. com had the top number of URLs delisted, with 7,712 and 6,682, respectively (Google Transparency Report, 2018).

The free flow of information may undergo compromise at the hands of a search engine data controller making a split-second decision to remove data. Communication performs a vital role in a democracy. The power resides with the individuals communicating and the ability for information and ideas to prevail in this Digital Age. Certain ideas will emerge from the monotony of ideas in the marketplace to garner public support and create a public opinion. Freedom of information flow is congruent with selection of political candidates and the spirited debate of political, economic, and social issues in a society. This chapter explores the First Amendment implications of the freedom of speech and freedom of the press within the context of cultural democracy while it contemplates the impact of the right to be forgotten.

Chapter 4

The Specter of Government Intervention

The specter of government intrusion into the free flow of information in society signals a loud alarm for many Americans who cherish the ability to freely communicate. Woven into the fabric of U.S. freedoms, freedom of speech represents a bedrock principle in a nation safeguarded and guaranteed by the U.S. Constitution and bolstered by years of successful case law inside federal courtrooms. The adjudication of First Amendment disputes in the United States and adherence to the principles of the Founding Fathers have given citizens expansive rights to communicate, so they freely speak with others and actively listen in a democracy. This chapter addresses the concept of government intervention, from both historic and contemporary perspectives, to explain the threats to the free flow of information advocated in various aspects of web governance. Further, this chapter answers the question why Americans bristle at the right to be forgotten, and the text dissects the threat of government intervention that paves the way for censorship, creates false realities, and bolsters the potential rise of totalitarianism. In addition, we explore the Absolutist Theory of the First Amendment as an aspiration to preclude government intervention. The idea of government intervention into what citizens can say and how they can say it presents a slippery slope to information control, a reality that residents of the United States don't readily endorse, opting instead for the free flow of information in society.

In light of the fact that the right to be forgotten represents government intervention in Europe, the worldwide information sphere is sculpted and changed because of the removal of online data, governance, and availability of information worldwide. Government intervention into the availability of information changes the way that journalists can perform their roles because it hinders the research process. These adaptations change the way readers can gain knowledge, alter public opinion, or adapt their perceptions for social, economic, or political reasons. Journalistic "prior restraints" and government removal of information from the World Wide Web typify activities of totalitarian governments. To unfold the circumstances presented today in a right to be forgotten environment, let's examine the presence of historical opposition to government intervention in public affairs in the United States, learn about philosophical insights set forth by philosopher Hannah Arendt, and explore the contemporary condition of information control that carries a censor-based, hegemonic practice from Europe to the United States. In consideration of these global implications though, we can further observe the impact of the right to be forgotten as incorporated by the European Union leads to changes in the application of law, stark moves that could result in unexpected consequences in the flow of information among European nations, the United States, and other democracies that acknowledge the importance of freedom of information in society.

The composition of American constitutional law considers the onus of government intervention into communication and repudiates it as a threat to the democracy. To better understand the foundation for this thinking, we can look to Colonial America, where early printers sought to present pamphlets and other printed materials to Colonial residents to inform them about government policies and collective resistance. The Founding Fathers witnessed firsthand the attempts by Great Britain to silence colonial voices through the use of required prior government approval, government stamps, taxation, licensing, and other measures to quell rising opposition to the Crown. For example, the British mandated prior approval of colonial press content and utilized seditious libel laws to punish government critics (Lewis, 2007). As suggested in chapter 1, contemporary beliefs still ring true for the [James] Madisonian point of view, suggesting that sovereignty lives within people (Sunstein, 1993). This notion resounds even more loudly today in light of the advent of user-based digital media. The power to convey information resides in the hands of the individual content producer sharing information via social media. Colonial printing presses possessed the ability to convey information to the new American colonies, while social media content producers today harvest the capacity to communicate with global audiences. Though citizens in various countries experience Internet regulation to varying degrees, the United States still ties its ability to communicate freely online to the bedrock principle of the First Amendment and any type of government intervention that would impinge upon the freedom of expression. United States citizens, bolstered by this strong journalistic heritage, call for the highest levels of transparency for government entities. The adherence to these transparency standards find their way into American laws, including the Government in Sunshine Act, allowing for the "sunshine" to enter public meetings where elected officials create laws and administer public funds, and the Freedom of

Information Act, mandating the disclosure of public documents with a few notable exceptions, which include documents related to national security and geological data.

While the contemporary dialogue of the right to be forgotten conjures the specters of censorship, government interference, and reduced transparency, citizens in the United States bristle at any type of movement designed to chip away the "power of the people" in the American democracy, even if that means sacrificing personal dimensions of privacy. Though the European Union may not exert information control using taxes and stamps, it can exert a far more tyrannical approach to information management: The European Union has emboldened private search engines operators as quasi-governmental censors. The predicament may be described:

> While the impact of the decision is couched in language boldly defending an individual's right to control what the public knows about him, it deleteriously casts Google and other similarly situated companies into the role of surrogate government censors, conscripted into service to rewrite history's permanent digital memory to omit unattractive facts that human memory has all but forgotten. (Locascio, 2015, p. 298)

The European regulation's right to be forgotten faces another practical challenge—enforcement. By empowering the private sector to enforce the act, remedies for noncompliance must be addressed (Kelly & Satola, 2017). Calling the actions of the data controller "uncertain," Sartor (2015) calls into question the ability of the search engine to honor all requests, and, if they fail, the extent of liability must be determined, along with the concepts of compensatory damages or administrative sanctions. Under the new regulations, in Europe the power now resides with the search engine, operated by a management team and specific individuals who receive the request forms and arbitrarily make decisions about matters they may not be trained for or qualified to properly manage. The mandate by the European Union to force Google to make a la carte determinations has been called "highly imperfect" because the private corporation is now thrust into a position to determine what is in the public interest, who is a public official, and how much of a public figure must they be to merit data retention (Locascio, 2015). Opposition suggests that a nongovernmental entity should not be given broad power to determine whether material is relevant or outdated, and sets the stage for patchy, biased histories (Peaslee, 2016). At first glance we must acknowledge that the inbound information to the search engine may be faulty, because the search engine forms rely on self-reported information from the respondent seeking to have information removed. This immediately disrupts the flow of information and the ability to vet the accuracy of the information. While the governments of the 28 countries in the European Union do not specifically request the change, they passively bestow this power onto a private corporation, creating an "arms-length" distance from the government, which essentially endorses modern-day censorship. The United States is a plane ride away from Europe and maintains commonly held views with citizens in the respective European democracies, but the interpretation of the European right to be forgotten places the United States and European countries on a collision course. The disparate interpretation of the search engine as a data controller in Europe varies greatly between the United States interpretation of the search engine and the corresponding generation of results. While the overriding force of the First Amendment guides discussion about the regulation, contiguous topics, including the regulation of algorithmic-based speech, the role of the Absolutist Theory of the First Amendment, and the viability of any workable alternative for the United States continue among legal scholars. Today, however, the winds of the First Amendment pervade the atmosphere, stifling movements from privacy activists and others that attempt to move the United States toward European censorship via search engine.

WHY UNITED STATES CITIZENS OPPOSE THE RIGHT TO BE FORGOTTEN

Advocates of the free flow of information in society react with alarm and fear that government-sanctioned censorship of the World Wide Web will lead to various social ills, including censorship, memory holes, and the re-writing of history. While First Amendment advocates can wave the U.S. flag to maintain free speech, several legal issues complicate the central argument. First, legal scholars debate whether search engine results merit First Amendment protection. Second, the speculative enforcement of the right to be forgotten in the United States generates legal issues surrounding the jurisdiction of the foreign country judgments in the United States. Despite the fact the European right to be forgotten may imperil free speech and create "memory holes" in the web, another question merits further exploration: Do search engine results earn First Amendment protection under United States law? This conjures up issues related to American jurisprudence and interpretation of constitutional issues. To illustrate this conundrum, various corollary questions emerge: How does the United States interpretation of the search engine result differ from the EU classification of Google as a data controller? How can these issues reconcile globally with differing views of information governance and territorial-based decisions?

In a 2011 paper to the European Commission, scholar Joris VanHoboken explored the nexus between the right to be forgotten and freedom of information, suggesting that tort laws and existing laws that govern media would provide a better framework to manage the freedom of expression concerns than the route of data protection. Further, in this early work he presciently established the groundwork that the attachment of liability to search engines as intermediaries conflicts with the intermediary safe harbors, and, if pressed, these intermediaries would invoke a media exception. This is exactly what happened in the Volokh and Falk paper commissioned by Google in 2012. The white paper from the UCLA School to Law explored the notion of whether or not the First Amendment protects search engine results (Volokh & Falk, 2012). The white paper takes a new tack that asserts Google and other search engines are speakers themselves who direct users to material created by others. Further, the dynamic selection of results accomplished through algorithms and editorial judgments may further solidify a First Amendment claim for protection over the search engine results. Following the logical trajectory that Google makes editorial judgments every millisecond, the authors point to precedence to bolster protection claims as the search engine rankings were "subjective results." Further, while citing a pinpointed path of jurisprudence, the authors conclude that the First Amendment clearly covers the search engine results. Through a legal analysis, Volokh & Falk (2012) underscore the notion that online speech is fully constitutionally protected, even when algorithms are used to organize data.

Do Search Engine Results Merit First Amendment Protection?

The patchwork of legal cases that courts have issued to guide legal considerations of search engine results addresses the use of opinion, the ability to "remain quiet," and the active role of search engines as editorial decision-makers. First Amendment protection for search engines erupted on the legal landscape when Search King, Inc., of Oklahoma sued Google Technology, Inc., asserting that Google purposefully and maliciously decreased the page rank of the company. The court found the page rank results represented an "opinion" and are thus entitled to First Amendment protection (*Search King v. Google Technology*, 2003 Case No. Civ-02–1457-M, W.D. Okla., Jan. 13, 2003). Later in 2007, the federal district court in Delaware held in *Christopher Langdon v. Google, Inc.*, that the First Amendment, in tandem with the Communications Decency Act, protects Google and Yahoo from refusing to run advertisements for two politically oriented websites they did not wish to run (*Christopher Langdon v. Google, Inc.*, *Yahoo Inc.*, and *Microsoft Corporation*, 2007, WL 530156, Civ. Act. No. 06–319-JJF). The case, decided in the United States District Court for the District of Delaware, looked to precedence to decipher the contemporary First Amendment understanding. The decision relied on a 1987 case (*Riley v. National Federation of the Blind of*

North Carolina, Inc. (487 U.S. 781, 796–97) to underscore the idea that First Amendment guarantees encompass the right to decide "what to say and what not to say," and reiterated the 1974 finding in *Miami Herald Publishing Co. v. Tornillo* (418 US 241, 256, 1974) that indicated forcing newspapers to print candidates' responses to editorials is an "impermissible" burden. Yet, the decision created a new frame by invoking section 230 of the Communications Decency Act, and, recollecting the decision in *Green v. AOL*, 318, F.3d 465 (3d. Circ. 2003), the court found that campaigns to assign liability to search engines directly counter their "immunity," derived from creating editorial decisions in the search engine.

This decision marked a new era of dialogue about algorithms that still captivates scholars and practitioners, as they wade through issues of First Amendment protection in the online search engine world. In attempts to yield First Amendment protection for the search engines, scholars have made arguments that assert the First Amendment should protect the speech rights of the search engine's constituents, that the search engine itself is protected because it exerts editorial control, and the search engine produces opinions, thus meriting First Amendment protection (Bracha, 2014). Scholars point to both sides of the argument when considering whether search engine results are classified as First Amendment speech (Minora, 2017). While the Volokh and Falk (2012) Google white paper asserted that the Google should gain First Amendment protection because the search engines exercise editorial control, this line of reasoning positions the search engine as a "speaker" that directs users to material created by others. The multiplicity of editorial judgments executed by Google every nanosecond and the fact that algorithms generate ranked ratings suggest the presence of First Amendment Protection, according to Volokh and Falk. However, the Communications Decency Act, a federal law enacted in 1998, "liberated" search engines from liability for indexed content (Minora, 2017). The debate about online search engine results erasure takes us headfirst in a debate of the creation and protection of algorithmic speech and the legal designation of search engines as data controllers. Algorithmic speech occurs when a technology or device uses an algorithm to create a verbal or written communique that may be classified as "speech."

Algorithmic Speech

It has been suggested First Amendment protection emerges when algorithms produce a message that can be conveyed to another, as was found in the Search King case. However, Wu (2013) asserts that functionality of "machine speech" must be considered. It is not so simple as to assume that algorithms can create speech that may be protected, but the function of the speech must also be factored into the mix. For example, do car alarms generated algorithmically merit First Amendment protection? Wu (2013) suggests that this type of thinking is faulty, when one assumes this type of algorithmic speech merits First Amendment protection. He suggests interpretation of the law may have gone awry when considering algorithmically based speech and attempts to close the gap by creating a line of distinction between a speech product and a communication tool. In a 2012 *New York Times* editorial, Wu writes: "The First Amendment has wandered far from its purposes when it is recruited to protect commercial automatons from regulatory scrutiny." In a later article Wu (2013) explores a "functionality doctrine" of the First Amendment, manifested by the fact that courts tend to withhold protection from carriers/conduits, and common functional tools such as navigational charts and the like, thus not granting typical "tools" with First Amendment protection. The legal justification for this may reside in the United States Communications Decency Act that states: No provider or user of an interactive computer service shall be treated as the publisher or speaker of any information provided by another information content provider (47 U.S.C. 230). The code's legal protections differentiate the United States from Europe, suggesting that the United States may be a safe haven for technologies that seek a legal environment that favors free expression (Electronic Frontier Foundation, 2017). New technology innovations will continue to test the boundaries of regulation. The creation of speech by artificial intelligence unearths new paradigms to be explored in the context of First Amendment regulation (Massaro, Norton & Kaminski, 2017). The role of the humanness in the communication process and the ability for humans to listen to expression created by artificial intelligence remain areas for future contemplation as we navigate freedom of expression in the Digital Age.

IS GOOGLE A DATA CONTROLLER? IF SO, WHAT DOES THAT REALLY MEAN IN THE U.S.?

If we consider the typewriter, the answering machine, Twitter, YouTube, Facebook, and other technologies as conduits for information, we must ask ourselves if global search engines are conduits or speakers? While conventional wisdom may classify the search engines as conduits, any move to censor the information would mean that First Amendment protection comes into play.

> Even if the communication in question would not otherwise be considered speech, a demonstrated censorial motive on the part of the government can trigger First Amendment analysis. In other words, a demonstrated purpose in

> restricting the content of speech on the part of the government can, just by itself, create First Amendment coverage. (Wu, 2013, 1514)

Wu asserts that a de facto functionality doctrine exists in First Amendment jurisprudence, bolstered by the fact, first, that courts tend to withhold protection from carriers and conduits and, second, that courts do not routinely reward tools with constitutional protection. In his legal analysis, Wu points to several First Amendment cases that suggest the state reserves some type of power over the communication processes, while it simultaneously protects the expressive actions of speech (Wu, 2013). Wu characterizes Google as a "highly advanced and powerful index," asserting functionality of the search engine would provide the division between protected speech and communication. Under the Wu model, in the application of the right to be forgotten, a censorial motive exists. In other words, we can affirm search engines are functional and might not be protected, but when legal intervention precipitates a censorial move, the First Amendment is triggered.

This argument does not gain traction without obstacle. The complicated nature of search engine legal regulation has been called a system with "many moving parts but few degrees of freedom" (Grimmelmann, 2007), and a multitude of competing policy demands. Some legal scholars oppose First Amendment protection for the algorithmically generated results. Bracha (2014) asserts the search engine rankings are not protectable speech, and Grimmelmann (2014) posits a better frame to consider search engines is to classify them as "advisors" rather than consider them as editors or a technological conduit. According to Grimmelmann, the editor theory ascribes editorial responsibilities to the search engine, while the conduit theory suggests the search engines are a "pass through" for information, thus relying on the First Amendment presumption of an affirmative right to access media (Grimmelmann, 2014). Instead, the advisor perspective emerges from the audience's perspective, through the trusted "Advisor" role whereby Google can promote access and loyalty. By attaching editorial judgment to the search engine, the First Amendment arguments come into focus. The notions of freedom of the press and editorial judgment are founded in the concept of broad distribution of information, and the exercise of editorial judgment is viewed as a core idea of freedom of the press (Bezanson, 1999).

Google's 2017 transparency report indicates the presence of human intervention when the company performs a "manual review" of submitted web forms and states: "There are no categories of requests that are automatically rejected by humans or by machines." The mere presence of human intervention indicates a form of "editorial control" outside the auspices of the algorithmically generated speech. The First Amendment protection of search engine results remains a formative discussion area. It will not be decided universally, but rather through a series of legal cases that emerge throughout the United States. We should not, however, lose sight of the larger issue in this case—the ability for citizens to discern actionable "truth" for social, economic, and political actions.

The Objectors—Wikipedia and the Google Legal Teams

From the outset of the European General Data Protection Regulation in Europe, "objectors" to the right to be forgotten have emerged, articulating central arguments about the First Amendment and underscoring the imperative for unbridled and uncensored information in a democracy. Specifically, Wikipedia's founder and Google legal teams set forth arguments against the transformation of search engines into data controllers. From the moment that Viviane Reding addressed the European Data Protection and Privacy conference in 2010 and called for more stringent data storage regulations, reflections from the global community have been mixed (Reding, 2010). At that time, she stated her objectives were to enhance the rights of the individuals to protect their personal data by reinforcing the data controller's responsibility. Yet, when introducing the concept of the "right to be forgotten" and citing social networks, she proposed an aspiration for individuals to have the right to wipe out their profiles, and, further, applicants should be able to exercise that same right after a designated storage period concludes. The British House of Lords repudiated the decision and stated that the interpretation of the search engine as a data controller was "misguided in principle and unworkable in practice" (House of Lords, 2014).

While the European General Data Protection Regulation carries on and Google continues to honor requests for data removal, outspoken advocates, including Wikipedia founder Jimmy Wales, Google's own legal team, and senior management at Google have articulated failings of the right to be forgotten. In a 2014 interview with *The Telegraph*, Wales called the right to be forgotten "deeply immoral" and warned against the Internet's future "riddled" with memory holes (Curtis & Philipson, 2014). Wikimedia's CEO Lila Tretikov warned the right to be forgotten sanctioned "censorship of truthful information" during a news conference in London in summer 2014 (Lomas, 2014). Further, the right to be forgotten counters the core values of Wikipedia, which provides citizens access to knowledge and the ability to contribute content. In the company's 2016 annual report for the Wikimedia Foundation, Executive Director Katherine Maher addresses the changing dynamics of the Internet—"The world is a very different place than when Wikipedia was started, the Internet is less open and more commercial. Privacy is increasingly rare. Sources of information are more fragmented and not always reliable" (Wikimedia, 2017).

First Amendment scholars represent the most outspoken opponents to the European right to be forgotten, including this book's author. Noted First Amendment advocate and constitutional lawyer Floyd Abrams articulates the common opposition: "It is not a small thing for a government, let alone a continent-wide governmental entity to criminalize the dissemination of truthful information on a medium that most people turn to for just such information. But that is now the law in Europe (Abrams, 2017, p. 76). Therefore, with the constitutional questions hanging in the balance, the doubt remains about the feasibility of any type of corresponding right to be forgotten in the United States. Could the United States incorporate a right to be forgotten?

A WORKABLE PARALLEL IN THE UNITED STATES?

Living and working in the Digital Age, citizens produce vast amounts of information about themselves and others. The popularity of social media channels has given rise to an unparalleled level of sharing intimate details, thoughts, and images of our daily lives. As information consumers we rely on Internet search engines to rapidly retrieve information we can use for a myriad of purposes, from finding a mate and investigating a job opportunity to understanding news issues or shopping for the latest fashions. So what do Americans really think about the possibility of a mechanism to take down old information and redeem their online persona? In the absence of hard data about the topic, researchers set out to poll Americans about their attitude to take-down requests, probing whether United States citizens think the task should be handled by a search engine or government entity; if support from the right to be forgotten is stronger, when considering the rights of children; and if criminal information should be excluded (Bode & Jones, 2017). The researchers found American citizens care about who will consider the take-down requests, preferring the search engines or websites themselves to have the power to remove the information. In this sense, researchers found private industry's removal of data is more palatable than government intrusion in the free flow of information (Bode & Jones, 2017). The research also suggests a demarcation of the right to be forgotten, suggesting that it takes three forms: either as the right to have information deleted after a certain time, the right to have a "clean slate," or the right to be connected to current information (Koops, 2011). An argument has also been asserted that an American right to be forgotten would limit the power of hackers if information could be de-linked and removed. Citing the notorious Ashley Madison hack of 2015 that spilled the names of users who signed up for the website uniting individuals seeking extramarital affairs, it has been suggested that the stolen data would have been more difficult to find if it has been de-listed (Stenning,

2016). In an ironic twist, the hack began because the hackers objected to the $19 fee for a "full delete" of registration and browsing history on the website.

The ability for an individual to exert control over an online persona remains central to this debate. Walker (2012) suggests the ability for an individual to delete information that they created and published may be salvageable as a form of a modified U.S. right to be forgotten that would potentially be compatible with U.S. Constitutional law. Privacy issues, First Amendment debates, and the assignment of responsibility in information control all circulate through common threads of argumentation, some asserting that an American right to be forgotten would serve society, with others suggesting that preclusion of the free flow of information clearly outweighs any benefits accrued. In light of the vast cultural differences between the United States and Europe, the value placed on privacy and freedom of information, the codification of laws, and enforcement of corresponding influences remain different. However, legal scholars suggest that, if a court must face a right to be forgotten legal issue, they must not enforce the ruling, because it contravenes the First Amendment, and the legal implications should be isolated to Europe because Europe's removal policies should not force the world to forget (Minora, 2017). Other scholars have suggested consideration of the right to be forgotten within the lens of United States defamation law and propose limiting the right to be forgotten to private individuals rather than public officials or public figures (Rustad & Kulevska, 2015). Instead, the United States should steer around constitutional issues by declining to adopt a parallel right but simultaneously should allow public dialogue about online privacy expectations (Royston, 2016).

While this text addresses the First Amendment challenges to a European right to be forgotten and explores corresponding issues related to government power, memory, and privacy, the question remains whether a parallel right to be forgotten can or should emerge in United States legal structures. The right to be forgotten legally addresses the enduring memory of the World Wide Web and establishes a tangible recourse for individuals seeking relief from the perpetual reminder of online information about them. The ability to apply for data removal typifies the process of "forgetting" that removes the data in question though search engine de-linking. While the European Union adopts the policies among the 28 countries in the Union, the United States continues to resist measures to incorporate similar provisions. Though First Amendment arguments take center stage for the opposition, movements to find a way through the forest of individual privacy rights, digital redemption, and contemporary global web governance march on. Legal scholars point to various U.S. laws that contain existing dimensions of forgiveness, including bankruptcy laws that allow for the discharge of debts, pardons, statutes of limitations, records expungement, and the orderly discharge of records in line with federal consumer reporting regulations (Ambrose, Friess, & Matre, 2012) as evidence points for potential legal footing for a parallel right to be forgotten in the United States. In particular, the U.S. Fair Credit and Reporting Act of 1970 (FCRA), designed to protect creditors from false or inaccurate information and prohibit entities from disseminating obsolete information, indicates a societal value to consumer protection and advocacy for the social value of forgiveness (Ambrose, Friess & Matre, 2012). Extending this rationale, it has been suggested search engines may qualify as credit reporting agencies under the law, because they regularly assemble or evaluate information on consumers for monetary fees (Andrus, 2016b). Google's parent company, Alphabet, Inc., indicated in the company's annual report the company earned $79 million in advertising revenue from Google advertising in 2016 (Alphabet, 2016). It has also been suggested that the United States create a parallel right to be forgotten by allowing individuals to remove information they have posted and provide for the ability to remove information that is offensive to a reasonable person or is not newsworthy (Owings, 2015). These efforts represent eager legal calisthenics that make attempts to decipher what is happening in the European data protection landscape and apply it to the United States legal code. The New York State Assembly received a draft bill in February 2017 that called for the right to be forgotten in the United States (New York State Assembly, 2017). The bill, introduced by David Welprin (D-Queens), has not gained traction for approval but the introduction indicates that United States lawmakers are contemplating the right to be forgotten.

When one considers the search engines as data controllers, a correlation between the European right to be forgotten and the American consumer's right to dispute emerges. The logic trajectory indicates consumers can "dispute" their personal data by virtue of the Fair Credit and Reporting Act, removing the inherent controversy of the right to be forgotten as a First Amendment threat. While the FCRA is posited as a potential solution, it is not the only remedy legal scholars have considered as they continue to scrutinize existing and speculative laws. While the First Amendment's broad protection for free speech would likely lead to the judicial rejection of any oppositional statutory construction, legal scholars imply narrowly tailored solutions may be appropriate. For example, in addition to the credit reporting dimension (Andrus, 2016a), scholars suggest that provisions for cyberbullying and revenge pornography should remain under consideration in the United States (Antani, 2015). To justify data removal, advocates often point to existing measures in the U.S. law, such credit reports deleting negative information after a set period of time, the expunging of juvenile criminal convictions, and the erasing of some misdemeanors (Pike, 2016). In recent years, the U.S. Congress has contemplated various legislation to enhance the rights of online users. Neither the Consumer Privacy Bill of Rights Act of 2015, which called for baseline protections of individual privacy, nor the Application Privacy, Protection, and Security (APPS) Act, which called for a new level of transparency in mobile applications, failed to gain broad support. However, the introduction of these bills that address relevant issues such as individual control and transparency, reveals a potential inclination toward elements of digital forgiveness (Stenning, 2016).

When it comes to the retrieval of online information, legal authors have floated various approaches to solving the dilemma of the perpetual eyes of the web, including: formulation of a legal state of "reputation bankruptcy"; the enactment of laws that preclude employers from utilizing online information about a person after a certain time period; decreasing accessibility by "expunging" search engine results in a similar fashion to the right to be forgotten; and utilizing "false light" claims under existing laws (Ambrose, Friess & Matre, 2012). In his assertions, Zittrain (2010) states reputation bankruptcy will not only give citizens a second or third chance similar to that found in personal financial bankruptcy and expungement, but will also provide a vital reason for citizens to be able to remove things from a "permanent digital record." However, it is possible that in the case of reputation bankruptcy, an individual would "throw out the good with the bad," paying the price for eradication of digital information and the subsequent fresh start.

A literature review suggests a potential legal framework that addresses the notion of forgiveness, encompassing prerequisites both for forgiveness and legal approaches. Scholars assert three factors must exist before seeking legal forgiveness: It must be a forgivable offense; a specific harm must have occurred; and a social benefit may accrue from the forgiveness. Further, the key elements of legal forgiveness must include facets of time, oversight, and relief from accountability (Ambrose, Friess & Matre, 2012). In an effort to find some aspects of digital redemption in the United States and reconcile the American position on the right to be forgotten, alternative models now emerge. Cook (2017) suggests a "flagging system," completed by the Federal Trade Commission, allowing for flagging when "private harms" outweigh the public interest. Stewart (2017) proposes a new test as a reaction to the extraterritorial reach of the European Union. The new test would consider the contemporary community standards, the presence of patently offensive material, and the value of information to a legitimate public interest. Further, Stewart calls for education among citizens to increase Internet responsibility and the creation of a well-defined bilateral treaty between the European Union and the United States. While each of these approaches offers noble gestures to the conundrum of online information preservation, they must ultimately pass First Amendment muster. Despite the free speech objections to the removal of information, recent developments in American law suggest potential bending of the rules to accommodate crime victims.

VICTIM RIGHTS AND REVENGE PORNOGRAPHY

While staunch advocates of the free flow of information guard against the preservation of online data, movements in the United States to protect crime victims and targets of revenge pornography indicate some evolution in the practice of what information remains indefinitely archived on the web and what information may undergo de-linking and removal. For example, a prevailing tenet among victims' rights activists is that a right to be forgotten in the United States is a justifiable privilege for citizens who find themselves as victims who did not purposefully or willingly become a participant. While debate surrounds the classification of crime victims as individuals with supplemental rights, the public's response to revenge pornography (the posting of sexually oriented content without the subject's approval) and the adoption of laws that ban revenge porn and provide victims with a legal recourse, suggest U.S. residents may have a threshold of tolerance for the right to be forgotten. According to the Cyber Civil Rights Initiative, by the summer of 2017, 38 states and the District of Columbia had adopted legislation to criminalize revenge pornography, with classifications ranging from misdemeanor to felony charges (Cyber Civil Rights Initiative, 2017).

To alleviate the pain of a memory of a past occurrence, victim advocates plead for a corresponding right to be forgotten that could apply to crime victims (Cooper, 2016). While the implementation of the right to be forgotten in the United States may not pass constitutional muster, because it is too broad and too vague, crime victims may have an additional need for legal protection resulting from hardship suffered as a result of a crime. Crime victim rights remain closely tied to victims' privacy rights, so without proper protection, citizens may fail to report crimes and thus imperil the criminal justice system and society (Cooper, 2016). "There is no reason why a right to free speech should outweigh a crime victim's privacy right, particularly because the individual is likely not a public figure and the victim clearly did not want to become a crime victim," Cooper writes (2016).

Yet, the victim and the perpetrator often remain linked in the content retrievable via an online search engine, making the de-linking of the story about the victim a practical removal of the story about the perpetrator. Information removal impedes the free flow of information and compromises the research capabilities of citizens and journalists gathering information about a crime. For crime victims, a suggested test considers information removal based on specific crimes, evidence of a direct connection between the crime and victim in the search engine results, and the use of a "reasonable person" standard to determine the legitimacy of a claim (Cooper, 2016). Under the proposed model, the search engines could exert autonomy or anonymity over the request process, but the courts would make final decisions in the event of a dispute. Further, it has been suggested that Google would not have to permanently de-link the information, but rather adapt the Page Rank algorithm so the information would appear after the first six pages of results (Cooper 2016).

Slipping Down an Information Slope: Lessons from Philosopher Hannah Arendt

In examining the decline of communication in twentieth century democracies, we may take a lesson from philosopher Hannah Arendt. Born in Germany in 1906, Arendt studied under well-regarded German thinkers, including Martin Heidegger, Karl Jaspers, and Rudolf Bultmann, and gained recognition for her breakout work, *The Origins of Totalitarianism*. She traced the roots of Stalinist Communism and Hitlerian Nazism to the destabilization of living conditions after World War I and to the Great Depression, which together gave rise to the confused people open to a single idea promulgated by the totalitarian regimes. Further, she theorized Stalinist Communism and Nazism were different from other forms of political tyranny and manifested themselves as a "novel form(s) of government" that injected terror and ideological fiction into rule (Arendt, OT, 1951). Arendt postulated that in totalitarianism, a "lying world order" takes hold, where lies replace reality. She advances this notion further by asserting totalitarian regimes differentiate themselves from tyrannical and dictatorial regimes by the "radical deception" found in the "lying" world order. Thus, for Arendt, the only remedy for totalitarianism and an "elated citizenry" is the existence of a "factual" truth (Birmingham, 2008). If, as Arendt claims, a "lying world order" typifies totalitarianism, we have an obligation to question government policies that pierce holes into reported truths, create false or partial facts, and rob the marketplace of ideas from a full picture of reality. For Arendt, the role of a factual truth is an imperative in the curtailment of totalitarian political structures. Considering these philosophical insights, we must contemplate the role of Google and other search engines when they exert arbitrary forces that curtail the availability of information from the citizenry.

The legal arguments suggesting the search engines deliver an "opinion" of desired results may gain further enlightenment by viewing the search engine regulation through an Arendt-oriented lens.

> No opinion is self-evident. In matters of opinion, but not in matters of truth, or thinking is truly discursive, running, as it were, from place to place, from one part of the world to another, through all kinds of conflicting views, until it finally ascends from these particularities to some impartial generality. (Arendt, BPF, 238)

Advancing Arendt's ideas further, Birmingham (2008) suggests that "bearing witness" to factual reality is the ground for the democratic public sphere and may function as a remedy to the elated citizenry deceived in the lying world order. By virtue of this thinking, we may contemplate that access to accurate and complete search engine results enable citizens to "bear witness" to factual reality and stem the rise of totalitarian forms of government. Yet, the right to be forgotten adjusts perspectives of reality through an adjusted prism. Therefore, when considering Arendt's notions in the context of the right to be forgotten, we observe that removal of data may deprive us of the particularities of information that may be required to build opinion and may lessen the ability of the citizenry to bear witness to the unfolding of social, economic, and political events. Further, the obfuscation of accurate information may warp perceptions of democracy itself, thus endangering the future of the United States, and European democracies.

While Arendt's notions of bearing witness have been applied to human suffering (Ivanovic, 2016) and human rights during statelessness (Lechte & Newman, 2012), digitally bearing witness must now enter contemporary conversations, as the Internet facilitates the global reach of individuals. The calamitous notion of the lying world order must cause citizens of the world to take pause, just as they have done when considering Arendt-style "lying world order" when contemplating the rhetoric of terror (Birmingham, 2007). Scholars have also perceived a deleterious effect of totalitarian propaganda, indicating that the nature of the monopolistic control leads to a gradual, yet profound, distrust by the general public for news and information (Friedrich, & Brzezinki, 1962). The importance of injecting Arendt and broader notions of information flow in totalitarianism into the right to be forgotten's discussion foreshadow that the de-linking and data-removal processes may, in fact, not only reduce the ability to bear witness but may also contribute to the accumulation of a lying world order that would leave citizens confused and desperate to determine truth in order to make decisions. So we must consider the philosophical streams of thought that suggest the right to be forgotten has the potential to foment totalitarian political structures and reduce the effectiveness of democracies by inhibiting the free flow of information. Today, we observe objections to the General Data Protection's right to be forgotten provision because of the global impact on information availability. Other misuses of communication technology may soon present themselves as major challenges to the First Amendment and information freedom in the United States. Today, the governance structures for global communication remain in the formative stages and governance of the World Wide Web will ultimately influence the way that citizens view democracy, social policies, and economic conditions.

HOW THE ABSOLUTIST THEORY OF THE FIRST AMENDMENT COUNTERS GOVERNMENT INTERVENTION

Arguments for search engine protection and regulation keep circling the First Amendment of the United States. The first clause of the amendment clearly states, "Congress shall make no law," suggesting that introduction and passage of laws designed to limit free speech are unconstitutional. The Absolutist Theory of the First Amendment has been interpreted to dictate that the propositions of the First Amendment are absolute and that government may adopt no laws whatsoever that abridge freedom of expression (Abrams, 2017). In the case of the European right to be forgotten, no law has been upheld on United States soil that influences the free flow of information, but instead a law created in Europe now affects the availability of information in the United States and in Europe. First Amendment concerns therefore outweigh the movements for increased privacy accomplished through government intervention in data publishing on search engines.

Government intervention places a powerful tool into the hands of the search engine, giving, in the case of one search engine, an amateur decision-maker prevailing power about content decisions. The introduction of government intervention into the search engine data can foretell an environment in which government requires private companies to alter data, surrender information, or eliminate unfavorable information. Journalists and media professionals who have long advocated for transparency will fall to the whimsy of untrained search engine content decision-makers. The U.S. Supreme Court has traditionally held that any law impinging on the First Amendment is subject to "strict scrutiny" that there must be a compelling government interest and the law must be narrowly tailored. The right to be forgotten may not likely pass the strict scrutiny test, as a compelling government interest may not be justified (Antani, 2015). Media professionals who adhere to the principles of transparency and the free flow of information may find the government intervention into media practices by way of search engine results not only as unpalatable, but also as a threat to society's information and its expression.

The Absolutist Theory of the First Amendment literally translates the meaning, "Congress shall make no law" abridging the freedom of speech and of the press to mean that no legislative body will create a law that precludes the First Amendment freedoms. On a tactical level, this approach suggests that municipal laws, congressional actions, and executive orders would be repudiated under the absolutist theory. The theory is the creation of Justice Hugo Black, who wrote the dissent in *Ginzberg v. United States* (1966). In light of a broad blanket of protection under the concept of absolutist theory, justices gave credence to a myriad of tests to determine the practicality for government regulation, drawing distinctions between commercial speech and political speech, protecting national security, and giving citizens recourse for maliciously false information through the tort of libel.

Looking Back to See Forward: Early Movements to Protect Web Content

Historically, it is important to understand the first indications of protection for online content under the law. As early as 1997, the Supreme Court began encountering cases related to the First Amendment protection of content on the web and found in *Reno v. ACLU* (521 U.S. 844, 1997) no basis for "qualifying" the level of First Amendment scrutiny applied to the web. In the Reno case, the 1996 Communications Decency Act was scrutinized for regulations of indecent communications in an effort to protect minors, but the CDA law was found to violate the First Amendment because it was a practical, content-based, blanket restriction of free speech. Efforts to repress freedom of speech on the web unearth many First Amendment questions, and the corresponding consideration of the right to be forgotten unveils a myriad of issues related to the legal definition of a search engine and the corresponding governance of these global enterprises that function as gatekeepers to information. The question of regulation becomes a query about search engine technology. Each disruptive era of mass media innovation, from the early days of radio to the advent of television networks that set the agenda for American dinner table topics, regulators have grappled with notions of regulation that harmonize with the First Amendment. The web today provides a global channel for communication and serves as the repository of information for generations. How can we as a global society formulate a remedy that lets Google off the hook for policing content and becoming a censor of information? We may find several clues by scrutinizing Supreme Court case law, but we should note the European Union's enactment of the European right to be forgotten thrusts the web, Google, and global citizens into the cowboy's "wild west" of information control, with competing agendas, incongruent implementations across the European Union, and continued pressure in the United States to reject overtures to regulate search engine content with de-linking.

U.S. Supreme Court case law suggests broad latitude for the First Amendment to cover information that is overtly stated in a publishing outlet. The *Miami Herald v. Tornillo* case (418 US 241, 256, 1974) in particular drew a line of integrity around the newspaper, underscoring editorial decision-making authority of the news organization. In *Pacific Gas & Electric v. Public Utilities Commission* (475 U.S. 1, 1986), conflict erupted over the distribution of a newsletter in monthly billing statements that could have included information that was contrary to the beliefs of the public utility. When a public interest group sought access to the billing statement to respond to messages provided by the utility company, the court found freedom of expression recognizes the right of the publisher to refuse to carry messages. Therefore, in light of this trajectory, if Google is considered a publisher of content, the company can unabashedly make decisions about the publication of materials. A third case that may enlighten this situation is *Hurley v. Irish-American Gay, Lesbian & Bisexual Group* (515 US 557), which involved the refusal of the South Boston Allied War Council to allow a group to participate in the annual St. Patrick's parade. The parade organizers asserted that forcing them to include the group violated their First Amendment rights—and the court agreed. The court found that requiring a group of private citizens who have organized the parade to include a message that they do not wish to convey violates their First Amendment rights. These cases indicate the complicating First Amendment variables of editorial control and access to various communication vehicles. Court cases will continue to advance through the legal system, testing the uninhibited nature of freedom of expression with government boundaries, web governance, and privacy issues. U.S. citizens who adhere to tenets of freedom of expression will press against the boundaries and stand on the United States Constitution's Bill of Rights.

The "New" Disruption: Dealing with the Infrastructure Problem

Each generation of human existence grapples with the advent of new technologies to disseminate news and information to citizens. We now remain at another crossroads, as we consider the exchange of global information and governance of an information system that traverses all geographic borders. Balkin (2012) has suggested that, while the First Amendment functions as an information policy for democracy, the struggle for freedom of speech rests on the infrastructure that is designed and implemented to handle the information. To preserve freedom of speech, Balkin calls for inclusion of democratic values into the infrastructure to protect democracy. For the Founding Fathers of the United States of America, inculcating freedom of information and preserving the democracy meant protecting the infrastructure of colonial printing presses and allowing for the unimpaired dissemination of materials.

Yet today, the information delivery infrastructure keeps changing shape, requiring new policy initiatives and points of view about freedom of speech in the hyperconnected era (Brock, 2016). This requires a renewed state of vigilance for the freedom of information as part of the democratic society. We may be entering a "new school of censorship that controls the digital infrastructure" rather than the "old school characterized by seizing media control and suppression of speech" (Balkin, 2012).

While legal tenets guide contemporary regulation, the First Amendment demonstrates an expansive malleability moving through time. Benjamin (2013) posits contemporary law may assert that algorithmic-based decisions, including search engine results, are not protected by the First Amendment, while it leaves other First Amendment jurisprudence in place. He suggests freedom of speech may aptly apply to core dimensions of political speech, to individuals rather than corporations, and contemporary interpretations must call for a greater focus on the audience and the "right to receive information." Discerning issues associated with the right to be forgotten provides topics of robust debate for new media scholars, legal analysts, and practitioners responsible for implementing the right to be forgotten. Administration of such rights in Europe remains as amorphous as the fluidity of the technology. To make sense of Google Spain, Google and its advisory team must march headlong into a global debate on web governance and the free flow of information in society, giving Google a much larger role in global citizenship than perhaps the technology company's management anticipated. Scholars forewarn the right to be forgotten portends seismic shifts in journalism to the "inconsistent" and "arbitrary" nature of the global search engines (Brock, 2016). The amorphous technology, evolving at a quickened pace, presents unparalleled challenges to those in the United States and abroad seeking to balance the freedom of expression with governance, thus confounding lawmakers and jurists. The European moves to allow de-linking and deletion indicate a greater danger in what Brock (2016) calls the "ratchet effect" of new laws that cast a shiver over freedom of expression.

Threats to free speech in democratic societies do not typically occur overnight, but rather chip away at fundamental liberties that ultimately contribute to the degradation of freedom of speech. The right to be forgotten, codified into European Law, remains flawed when seen through the lens of global information exchange. Despite well-intentioned efforts to allow for individual freedom and privacy, the law encounters shaky ground when it crosses the Atlantic to the United States. Constitutional questions erupt, as the First Amendment comes to the focal point. Europe's approach to classifying search engines as data controllers means we must now consider whether search engines exert editorial control over search engine results and if the search engine results justify protected speech. We must ask, "What is the role and responsibility of the private corporation operating in a global economy of information exchange?" The right to be forgotten is an interim solution for ever-evolving governance, a fluid right residing on legal assumptions that the search engine functions as a data controller. Yet, consideration of the global implications foretells a chilling atmosphere of potential information suppression that would intervene with social, economic, and democratic happenings.

The implications of the European Union's actions cast a wide net, as they may impact global trade (Kelly & Satola, 2017), and the evidence of government intrusion into the marketplace of ideas paves the way for censorship and the rise of totalitarianism. Efforts to match the right to be forgotten puts the United States on a slippery slope of potential government intervention that presents worrisome implications for journalists and advocates of the free flow of information. While governments may attempt to balance "rights" or lift privacy concerns above the First Amendment, we must constantly guard against the specter of government intervention into freedom of speech and the press in a democratic society. The removal of data also contributes to the dangerous result of distorting history and creating memory holes.

Chapter 5

Distorting History

Memory Holes and Other Destructive Forces

Students in the United States engaged in high school and college activities receive repeated advice from mentors and teachers that "the Internet never forgets," as a warning to exercise caution when placing personal data onto the web and social media. Pictures of fraternity brothers drinking beer in a "keg stand" or a scantily dressed young woman having a blast with friends at a bachelorette party remain housed on servers of social media networks, creating risks for students after they graduate from college and attempt to get their first job. A Google search of the individual's name can render these images (and more) about the job candidate, providing intimate details about the person's activities, judgment, and ethics. Some may argue whether or not the transgressions of a 20-year-old should be allowed to cast light on the career of an older, more mature individual. Despite the discomfort these images may provide for the individual, citizens in the United States today cannot simply make a formal request to remove data for the deletion of historical information posted in bad taste or with unsound mind. The ability to recollect all of the details of one's past presents us with a wide spectrum of emotions and can serve up digital post-traumatic stress disorder, if the memory is exceedingly painful, according to columnist Hill (2011). This conundrum generates a healthy dialogue about web governance and the issues associated with online rights and reputation management.

The classification of European search engines as data controllers has thrust open the door for arbitrary censorship of information and influences the global availability of data. This chapter explores the underbelly of the distortion that could occur by analysis of the impact of memory holes and other destructive forces during contemporary information retrieval and curation. This chapter includes a discussion of power within the context of information control. The changing use of technology in our culture indicates a shift to constant "remembering" rather than experiencing the virtue and blank slate of "forgetting." Where does that leave society as we contemplate citizens living and working in a nation where the web never forgets and few recourses exist for de-linking of information that may be embarrassing or inappropriate? The ability of the search engine to instantly retrieve our proud accomplishments and regretful embarrassments in nanoseconds suggests global society must contemplate a new way of framing memory and how we share, preserve, and hold memories of the past.

In a philosophical pursuit of the changing dynamics of living and working in the Information Age, scholars have attempted to examine the social concepts of "forgiveness" and "forgetfulness" as redeeming social values that enable an individual to harvest a fresh start (Blanchette & Johnson, 2002; Ambrose, Friess & Matre, 2012). The new landscape encompasses a greater quantity of data available about individuals, with deep granularity about transactions and activities that can be correlated and utilized for predictive analytics (Blanchette & Johnson, 2002), contributing to conditions whereby social memory is digitally enhanced to never forget. The increase in availability of data has prompted some scholars to recollect English philosopher and social theorist Bentham's institutional building called a Panopticon as an analogy of the enduring watchful eye. Bentham (1843) set forth the notion that a building's architectural structure could allow for the monitoring of all inmates by one watcher without the inmates able to determine if they are being watched. While the watcher is not physically able to watch all of the residents simultaneously, inmates must act as though they are under constant surveillance because they do not know whether or not they are being watched. This notion was further invoked by French philosopher Michel Foucault as an analogy for contemporary societies that perform constant social surveillance leading to a disciplinary society and social quarantine (Foucault, 1991). With the digitally enhanced surveillance capability ubiquitous in society today, some suggest a contemporary imperative for privacy dimensions. To avoid the digital "panoptic society," one characterized by a perpetual watchful gaze fueled by the web and search engine results, researchers suggest data retention issues need to be a part of a personal privacy comprehensive policy (Blanchette & Johnson, 2002). While acknowledging censorship occurs when the right be forgotten requests began on the web, contemporary scholars now draw distinctions between various types of censorship in the Digital Age, classifying censorship of original sources in information storage; censorship of initial encoding of information; and censorship during information retrieval (Gorzeman & Korenjof, 2017).

The right to be forgotten remains situated in a larger discussion about web governance and data retention, yet the right to petition a search engine for removal of data remains the underlying anchor that confounds the United

States today. The right to be forgotten has also been called the right to oblivion or *droit á l'oubli*. Today, First Amendment advocates take a staunch approach to deter any forms of censorship, while privacy activists strive to chisel away at constitutional protections by citing issues of personal privacy, the advent of self-censorship, and the intellectual notion of societal forgiveness.

DISTORTING HISTORY

The ability for online search engines to remove information from the online tapestry of data available for public consumption can warp public opinion and change the way individuals perceive information. Historians like De Baets (2016) cast the right to be forgotten as a rewriting of history that "impoverishes our insights" into anecdotal lives and larger historical trends. By itemizing the distinctions between the right to forget and the right to be forgotten, De Baets reveals the differences of the legal heritage of the concepts, the former derived from a right to freedom of expression and the latter from privacy.

Like two sides of a coin, the right to forget includes a complementary right to remember, consistent with the healthy exercise of freedom of expression. In contrast, the right to be forgotten presents a "radical variant" of the right to privacy, according to De Baets (2016), and suggests an element of coercion, because the person seeking the right to be forgotten diminishes the right of information for others. Globally, Article 19 of the United Nations Declaration of Human Rights states: "Everyone has the right to freedom of opinion and expression; this right includes freedom to hold opinions without interference and to seek, receive, and impart information and ideas through any media and regardless of frontiers" (2016). For scholars like De Baets, the coercive element of removing the information becomes the linchpin for understanding that the "right to oblivion" is in fact an infringement of freedom of expression. The dangers of information removal present society with new considerations for preservation of history and collective memory.

Memory Holes

Obliterating information establishes a dangerous precedent for society, suggesting new aspects of power for government and private enterprise, reminiscent of a dystopian society. In George Orwell's novel, *1984*, the term memory hole appeared to describe government officials creating slits in the wall to discard massive amounts of documents by hurling them through pneumatic tubes to a blazing fire. The documents in question conflicted with the prevailing regime, and the sequential wiping information away from history ripped a memory hole into the society's context and understanding of contemporary happenings. In a lament of the decline of memory, "Everything faded away into a shadow-world in which, finally even the date of the year had become uncertain" (Orwell, 1949, p. 41). In this murky circumstance, the totalitarian government utilized a Ministry of Truth to reconstruct the past and generate propaganda in multiple media forms to disseminate to the citizens. Orwell's tale, often invoked when addressing revisionist history or censorship, can peel back the layers of philosophy that asserts that the right to be forgotten threatens the integrity of information found online because it creates online "memory holes" that forever change history and corrupt the interpretation of the Internet as a haven of preserved information (Antani, 2015). The Orwellian protagonist Winston Smith laments, "When there were no external records that you could refer to, even the outline of your own life lost its sharpness" (Orwell, 1949, p. 32). The availability of government stored data, information retrieval, and the erasure of information remain issues that government officials must address on a local and global level.

Thus, we encounter a double-edged sword in the Digital Age when considering memory. Do we want an authoritative memory that becomes archive-like and permanently assigns all attributes to us—from our fashion mistakes to a thoughtless failure that we have since grown beyond—and seek to more appropriately manage our own personal brand? Yet, some scholars argue citizens should have a right to remove undesired information from the web's infallible memory and contend stored information may no longer serve a purpose, or may, in effect, cause societal damage (Mayer-Schönberger, 2009). Some philosophers posit we are moving to what has been called a new memory ecology (Hoskins, 2011), one that considers the influence of media on memory. In this concept, we no longer have to rely on our personal memories or static photo albums and scrapbooks to recollect the days of the past, but now we have moving images and can "google" long lost friends to determine whatever become of them and were they successful? In trouble with the law? Happy in their family life? Even obituary notices travel across the miles with warp speed to bring the past into focus in the present and constantly remind us of what has happened. While the memory in our mind may dissipate the colorful nature of memories as time passes, the web rapidly recollects them in a contemporary mosaic of what makes up an individual's past.

TROUBLING TALES OF TOTALITARIANISM: ARENDT ON HISTORIOGRAPHY

After the discussion in chapter 4 about the advent of "a lying world order" addressed by philosopher Hannah Arendt, we can investigate her work,

Origins of Totalitarianism, for further insights to explain the trouble that may arise when a nation state does not have an accurate, documented history. Arendt warns against the negative results of failing to secure an accurate history when a totalitarian regime takes hold. To advance the rhetoric and power of the totalitarian regime, the very nature of objective historiography is punished and damaged beyond recognition:

> To this aversion of the intellectual elite for official historiography, to its conviction that history, which was a forgery anyway, might as well be the playground of crackpots, must be added the terrible, demoralizing fascination in a possibility that gigantic lies and monstrous falsehoods can eventually be established as unquestioned facts, that many may be free to change his own past at will, and that the difference between truth and falsehood may cease to be objective and become a mere matter of power and cleverness, of pressure and infinite repetition. (Arendt, 1951, p. 333)

The notion of historiographic forgery haunts historians and citizens concerned with the accurate preservation of information. The powerful forces that control the future by enforcing beliefs also possess the power to interpret, falsify, and control the past. This must be considered in the context of the availability of information on the web. It has been suggested that, in contrast, an accurate archive can counterbalance totalitarian manipulation (Finigan, 2011). By looking at two dystopian works of literature, Orwell's *1984* and Margaret Atwood's *The Handmaid's Tale*, Finigan demonstrates how the manipulation of the archive gives way to confusion and the rise of power. As a reaction to these cataclysmic dimensions of falling to a totalitarian regime, a passion for the "right archive" may emerge. The preservation of the "right archive" then could potentially curtail totalitarian manipulation of concepts, but it could also be argued that the archive itself could be used against individuals, families, and subversive organizers to quell any threats to the government. While totalitarian governments utilize propaganda to persuade the masses and debunk opposition to social, economic, and political policies, the archive itself can poke holes into the propaganda and provide an invaluable tool of contemporary journalism and citizen fact-finding. This phenomenon is precisely why totalitarian regimes filter content available to citizens and suspend the availability of online information for citizens. The right to be forgotten represents a Digital Age dilemma that may portend a practical rewriting of history as vile as the suspension of accurate information found in the Hitler and Stalin regimes.

As we consider whether or not search engines are considered Digital Age archives, it is relevant to further explore the philosophical heritage and corresponding implications of the archive. We may look first to French philosopher Jacques Derrida, who presents readers with the notion of the archive in traditional Freudian terms, particularly in connection with a "death drive." By examining Derrida's "archive fever" more closely, we can observe Derrida's position that the archive can be a historical point of view, but the supposition that archive fever is relevant to the digital creations attaches new dimension to our role when we consider the advent of citizen communicators.

A GLIMPSE INTO MEMORY STUDIES

The cornucopia of historical content available with just a few keystrokes ushers in a new virtual "public stage," where citizens increasingly struggle to define their reputation and protect the integrity of private information. So the question then becomes: How do Digital Age citizens safeguard their online personas? The ability to sculpt one's own persona becomes particularly challenging when others actively contribute content and display images 24 hours a day, seven days a week, and 365 days a year. The web hums with stories created by one's self and by others—composing a persona not only derived by the individual, but also by the online community. In contemplation of this contemporary situation, it has been proposed that the right to be forgotten does not distinctly represent an infringement on the right of expression nor does it function as a privacy guarantee, but the right to be forgotten establishes an environment whereby an individual may aptly construct a personal, though revisionist, narrative. By enfranchising the individual this way, the right to be forgotten matures into a societal-level concern about society's memory and memory rights (Tirosh, 2017).

Human memory is a complicated facet of the human condition. Scholar Daniel Schacter reminds us humans assign consequential bias to their memories:

> We extract key elements from our experiences and store them. We then recreate or construct our experiences rather than retrieve copies of them. Sometimes in the process of reconstructing we add on feelings, beliefs or even knowledge we obtained after an experience. In other words, we bias our memories of the past by attributing them to emotions or knowledge we acquired after the event. (Schacter, 2001, p. 9)

In the book, *Seven Sins of Memory* (2001), Harvard University professor Schacter points to the human memory's malfunctions calling them "sins," including transience, absent-mindedness, blocking, misattribution, suggestibility, bias, and persistence. However, these sins of memory resemble human qualities, not those found in algorithmically generated memory found in online search engines. Through memory, humans have an ability to self-regulate the historical onslaught of information that would be too painful to bear, but the digital memory offers no latitude of forgiveness—rather it returns the unpleasant, the untoward, and the deplorable to the surface with just a few keystrokes. Therefore, in the Digital Age we become reduced to the conclusions that can be drawn about our person from our online search engine results.

Who we "are" online defines us without any method to remove unpalatable information, and these depictions become a reputational concern. For Europeans, the obstruction to preserving one's reputation was exceedingly repugnant, paving the way for the acceptance of the right to be forgotten. The United States perspective on the reputational dimensions is somewhat different, as America espouses a culture of transparency that does not generate barriers between the private and public spheres. To counteract movements for information removal, the strategic act of writing more and speaking more about one's self on a particular topic provides an alternative that allows the individual to harvest a certain gradient of control over information about themselves (Janes, 2016). The right to be forgotten occurs at a time when residents of the United States clamor for greater transparency in institutions. Americans urge transparency in business, seeking to have corporate executive management comply with federal regulations like the Sarbanes-Oxley law and provide authentic, accountable quarterly and annual financial reports. Pop psychology underscores transparency in personal affairs by advocating open avenues for discussion; citizens call for transparency in accounting; and politicians suggest that the ability to shed light into a situation creates a higher level of accountability.

As individuals though, many "webizens" want to simultaneously see and be seen in the online environment, until something gains the light of day that does not cast them in the most favorable light. Today, we live in a transitory time in our global digital history, an era when decisions about information handling, storage, and recall will influence our world for generations to come. The confluence of legal principles with digital operations, cultural mores, and individual beliefs creates an international quandary for those responsible for implementing global information policies, most notably the right to be forgotten. The connection between media governance and cultural memory remains undeniable. Some scholars assert the Spanish court's decision and the institutionalization of the concept into European law denote the beginning of a new era that links media governance and regulation to society's information processes (Tirosh, 2017). Further, some scholars indicate that the right to be forgotten contributes to contemporary scholarship on memory, because it legitimizes the governance of personal and collective memories (Pereira, Ghezzi & Vesnic-Alujevnic, 2014). In an examination of archival practices, researchers say that the right to be forgotten will not completely harm preservation (Dulong de Rosnay & Guadamuz, 2016). The pair assert that the right to be forgotten is actually a misleading term, because the de-linking that occurs after the individual's request does not permanently remove the item from the original source, thus not promoting a total "disappearance" of the data. The concept of the archive is addressed below, but digital citizens have come to rely on Google as a personal archive, making the mission to find information more onerous by tracking down individual sources. This can be found in contemporary news organizations in which the institutions preserve their own archives, but when Google de-links them, the news item, though remaining covertly in the news archive, becomes less accessible on the search engine. This places an additional burden on information gatherers, including journalists or common citizens seeking information to help them make decisions.

As global citizens march into a digital era where items cannot so easily be forgotten, it has been suggested that we are entering a "post-scarcity" culture whereby the ability to shape one's online reputation and persona becomes more viable, yet simultaneously unwieldly, because of the continued return of information (Hoskins, 2014). The scarcity culture of days gone by, articulated in the mass media era of Marshall McLuhan and the prevalence of primetime media, constantly creating content that was "fixed" in time, stand in stark contrast to the media of today that is malleable, based on the user's whim, and morphs through the transmission of online media, giving social media greater power to depict the human being. In what has been called the "circulation of the self," the individual, culture, and society are reconstituted in real time, creating a digital "present" (Hoskins, 2014). Yet the right to be forgotten stands as a "re-sentiment" of the post-scarcity age, countering a memorial force that is uncertain about the prospects for erasing or forgetting (Hoskins, 2014). Forgetting and remembering represent the dual functions of memory, and the notions of individual and collective memory cross multiple disciplines, including psychology, sociology, anthropology, history, and communication. The classification of European search engines as data controllers has thrust open the door for the arbitrary censorship of information and concepts on a global level. As citizens living in a democracy, we must explore the dark side of the distortion that could occur, analyzing the impact of memory holes and other destructive forces in information curating.

INTERNET ARCHIVES AND THE LINK TO BIG DATA

With the preservation of images and information online, individuals become frozen in time with the presentation of data, whether outdated or unavailable for viewing. As humans, we conduct searches about other individuals, whether screening an employee or gaining background information about a potential romantic partner. The constant gaze of the Internet archive may not effectively allow for individuals to change, evolve, or be "different" than they once were a decade ago, a year ago, or a month ago. Cheng (2016) suggests de-linking and removal of the information is the wrong path, opting instead for an environment that allows for updating but does not counter freedom of expression. By creating a "right of reply" and facilitating contextualism, Cheng suggests a "right to be different," thus altering the concept of the person frozen in time. It has also been suggested that instead of contemplating the binary thinking of "forgetting" vs. "remembering," alternative methods could be employed, including "re-ordering" instead of de-listing (deMars & Callaghan, 2016). These lines of query may be the most fertile for individuals attempting to find a way to incorporate a similar right to be forgotten in the United States.

In sum, however, the movements toward web governance must appropriately consider the future of preservation and retrieval of vast amounts of data. Challenges for regulating the web archive and retrievability of information remain intertwined with the complicating notions of Big Data practices. By analyzing the right to be forgotten through the Big Data prism, Koops (2011) suggests three speculative frameworks for a semblance of digital forgetting. The first approach encompasses the idea that personal data should be deleted in "due time"; the second approach adheres to a social "clean slate" that diminishes the data used against a person; and the third perspective relinquishes social stigmas by allowing people to express themselves in online media. The absence of consensus and the ability for private corporation and government regulators to come to grips with ethical management of the information leaves Digital Age citizens adrift, fearful of how their personal information will be used and concerned about the misappropriation of their personal information that will in some way harm them.

The clamor for access to the digital archive and the robust memory of the web resides in a general need to understand the other. As a society, we are seeking to know *who* is that person we work with, how can I learn more about that person's habits, personal life, hobbies, and what may we have in common that could bind us together? We have a longing to engage with the other in the present sense. While we are all citizens, some have achieved a new social and legal status—that of a public figure with profound name recognition, to a public official elected by the people to serve the public. Some scholars have attempted to decipher this classification as "public" or "private" individuals, delineating their ability to request removal of information. In a laudable attempt to apply the body of U.S. libel law to the right to be forgotten, as mentioned in chapter 4, Rustad and Kulevska (2015) suggested a potential framework to deal with these different classifications of people and the right to be forgotten. The proposals expand on "three degrees of deletion," including the first for a data subject's own postings and pictures; the second for information that a data subject posts and third party re-posts on the third party's own website; and the third when third-party-post data is not created by the data subject. By uniting the landmark media law case, *New York Times vs. Sullivan*, that created a higher burden of proof for public officials to gain libel relief, the pair suggest a schema they indicate strikes a "fair balance" between the "right for reputational reset" for data subjects and the right to freedom of expression. However, if we recollect the marketplace of ideas, we may recognize that the individual may migrate from a position of private citizen to a public official or to a public figure during the course of a lifetime, and, as fellow voters, we may seek access to the digital archive to peer into their likes and dislikes to determine if we want to vote for the individual and endorse them to represent us in a public body. In an effort to outmaneuver the search engines, individuals who want to manage their online reputation now resort to third-party services such as Reputation VIP, which perform search engine optimization services for private individuals. In addition, European residents who want to leverage the right to be forgotten can tap into a centralized service at forget.me, which aids in multiple requests.

Collective memory is addressed by scholars in multiple fields, including those in sociology, history, psychology, medicine, and other areas. One of the most intriguing interpretations when contemplating the role of the search engine and the right to be forgotten is the discussion that collective memory is both a cognitive and social process (Casey & Olivera, 2011), and collective memory emerges from material artifacts and collective consciousness (Fine & Biem, 2007). This idea is advanced when considering that sociomaterial traces of memory are stored in archives, allowing for future retrieval and analysis (Ocasio, Mauskapf, & Steele, 2016), and ultimately a metanarrative emerges from the logics that emerge from the archives. For the sake of argument, the search engines of today consequently become a vital instrument for shaping of the metanarrative by determining the ranking, availability, scope, and societal nuances attached to information retrieved after a search engine query. But, the power of the search engine transcends beyond the ability to color the world by painting search engine results in a very specific way. The search engines also possess the power to discern social, economic, and political concerns of citizens, based on their search engine history, and can use that information for material gains, including financial, political, or social advancements.

The search engines themselves aggregate vast data warehouses full of data. The aggregate data, when interpreted collectively, can decipher topics of interest to Americans living in a certain part of the country. Topics of interest could include political leanings, public opinion on social issues, or a myriad of other conclusions based on information, not simple hunches. It is this thinking that led former Google data scientist and *New York Times* writer Seth Stephens-Davidowitz (2017) to explore the disconnects between Google search history, attitudes, and behaviors. Coined "the digital truth serum," he explores the "truth" about topics such as sex, prejudice, child abuse, abortion, and other social topics, finding dissimilarities between the language used to describe things in a social media environment and the language used in search engine searches. He cites the way people describe their spouses on social media sites as "the best, my best friend, amazing," but in anonymous searches use words like "gay, a jerk and amazing" emerge.

By examining this data, researchers can now surmise what is on the societal "collective mind," allowing the web to serve as an associative memory for users (Poritz, 2007). Yet this "collective mind" created through what has been called a search engine oligarchy (Poritz, 2007) derives value and creates nervousness among individual data specimens. For example, Poritz states that netizens may not understand that simply entering a single word into a search engine could help another person or corporation derive a strategy for investments, politics, or marketing. But, in that reality, the knowledge is what creates the squeamishness in collective privacy. The formation of subsequent laws such as existing prohibitions against insider trading may expand in the future to address these privacy concerns (Poritz, 2007).

DERRIDEAN ARCHIVE FEVER IN THE DIGITAL AGE

As we further examine memory and the advent of archive, we may look to Jacques Derrida's "Mal d'Archive" or *Archive Fever: Freudian Impression* (1995), because the author deconstructs the word root *arkhē*, to invoke a double signal of what Derrida calls the commencement and the commandment. The *commencement* dimension suggests a place where things of nature and history begin, while the *commandment* dimension is a place where authority and social order are organized. The duality of the underlying word affirms the deconstruction of the archive concept (Rand, 2010). By bringing this archive concept forward to the Digital Age, Derrida's suggestion of the dual nature of the archive seems appropriate, when we navigate the complexities of a digital archive found in repositories of Google, Bing, and other online search engines. From a commencement perspective, the digital archive represents a place where individuals and cultures store and retrieve data. From the commandment aspect, as global citizens we now personally experience how the global order of information remains in flux, because the European Union enacts the Data Protection Regulation encompassing the right to be forgotten and the United States legal and First Amendment scholars attempt to rebuff efforts that some claim change the free flow of information in our society.

To further understand the digital archive, we must first comprehend the role of both the power systems and the individuals in the creation of the archive. It has been asserted that *arkhē* gains its importance through signification in that the archive depicts a conceptualization of how knowledge is considered in culturally specific and ad hoc situations (Timcke, 2017). Further, while the archive presents society with a point of order and a documentable history, that at any time can be argued, the Derridean archive denotes a place where the objectification of memory occurs and this "place" is outside the human mind and human body.

It has been argued the desired signatures created in the archive represent a trace of human memory, articulating a shared, lived experience for human beings at a particular time in history. Timcke pushes the Derridean notion further by arguing that by virtue of participation with the archive, the humans seek to preserve memory in an exterior sense, suggesting that we do so in our struggle against death to validate a temporal existence by registering a presence in the archive (Timcke, 2017). Let's take, for example, the contemporary digital archive found at www.YouTube.com. This repository of moving images contains content created by millions of individuals. The motivation of posting these videos may include the ability to create a video to sell a product, share an idea, teach an audience about a particular belief, or any combination thereof. The motivation of creating these videos is as rich and broad as the diverse videos themselves. Despite all of the varying reasons and video iterations uploaded each day, they all have one commonality—they leave a trace. Through the posting in the digital archive, the video becomes entered into a channel, indexed into the system, and retrievable through the search engine, creating a sense of permanence. This permanence reflects the individual's image and suggests attributes of that individual's reputation at that time.

We now must consider what happens to that trace of memory when it is "damaged" or eradicated by a de-linking process that virtually "erases" the memory of the video's existence and alters the availability of the trace. To further complicate the matter, we must now additionally consider that the digital archive is algorithmically created, perhaps out of touch with the cultural knowledge that a human-generated archive would transfer into archival results. In a 2012 article, Boshoff points to Levinas's use of the term "an-archy" referring to the state that precedes the *arkh'e*, suggesting a primitive state before the commandment phase. From a deeply buried note at the end of *Otherwise than Being*, we can glean a greater understanding that anarchy cannot be sovereign, but disorder has an irreducible meaning, as a refusal of synthesis (Levinas, 1974, p. 194). The purposeful creation of disorder by disrupting the search engine algorithmic hum suggests the possibility of an irreducible meaning. We have to ask ourselves if holes in the search engine, the de facto digital archive, create an an-archy? Does this create an information anarchy, a place where disorder reigns supreme and contravenes the formation of an individual and societal sense of order?

While some may consider this overreaching or a dystopian reality of the Digital Age, the control of online information is a condition that we face as competing forces try to "control" what we see and when we view information. Google has its algorithms and takedown form, the European General Data Protection Regulation requires the search engines to comply with scarce recommendations for determining what is outdated and no longer relevant, and critics assert the absence of transparency makes the whole right to be forgotten a confounding circumstance. What will happen to our collective memory? While we may lament the dismal failure of the digital archive to create a proper concept for moving images, photographs, and the persona of an individual, the reputation of a corporation, or the progression of a social movement, we are not likely to forget the circumstances considering their continual return on our smart phone screens, tablets, and computers.

WHEN WE NO LONGER FORGET

The delicate nature of memory has been called a "blessing bestowed by the gods" (Schacter, 2001), and the mysterious nature of what we remember and what we forget provides psychologists and neuroscientists with an enduring challenge to understand how and why the human being remembers. The Digital Age now provides a new tool of memory—a hyper-vigilant device of aided recall that can bring forth the past in nanoseconds, thus denoting an evolutionary epoch for humans, as we adapt from a society that easily accepted the notion of forgetting, to a society that can no longer deny the unforgettable nature of information and facts that spew forth from the search engine results. As we swallow the implications of continual memory, we must pay homage to the fact that society's that ability to forget has been woven into our collective consciousness for generations. Victor Mayer-Schönberger's book, *Delete* (2009), documents the role of forgetting in society, characterizing forgetting as a condition required to help individuals make pro-active choices unencumbered by the past. The Internet Age has transformed societal forgetting from a natural state of losing information to information perpetuity, giving data a permanent longevity and providing citizens who live in the United States with access to a comprehensive digital archive that can contextualize an individual, for better or for worse. The maturation of tools of external memory, encompassing everything from books and video cassettes, audio recordings, podcasts to compact discs and online digital archiving of all of the aforementioned media formats, has turned the tables on forgetting. Remembering was once a rare state, as human memories failed or "wanted" to remember only selective elements of an event. Now, forgetting presents the rarity, because the digital archive routinely resurrects a wealth of information about an individual, forever morphing the ability of the individual to craft his or her own persona by virtue of the fact that external contributors can post data and tag individual photos. The digital revolution hearkens a new era of remembrance that some praise for the ability of individuals to aid simpler personal recollections to the expansive ability for governments to discern policy decisions (Burkert, 2007).

The shift to the perpetual gaze of the archive presents challenges for contemporary netizens. Mayer-Shönberger (2009) calls society "naïve" to think that human beings could easily leave behind the fundamental world of forgetting without pain. We may consider the right to be forgotten a growing pain in the maturation of data preservation in the Internet archive. The pain is experienced by contemporary society, both in the United States and abroad, and the transition to an all-remembering society may take generations to absorb. The threshold of tolerance for information preservation remains high as the generation known as the Millennials, who were generally born between 1984 and 2004, do not as easily recall the era of "forgetting" that their parents and grandparents once new, but the Millennials have grown up under the constant surveillance of the digital eye, just as their parents dutifully documented every achievement, smile, or frown, new pet adoption, and a few embarrassing moments along the way. The loss of forgetting in society, though, influences not only the individual, but also the societal structures on which the individual engages. To illustrate this point, Mayer-Schönberger (2009) points to three prongs in the shift in information power resulting from this loss: new-found aspects of information accessibility; information durability; and sheer comprehensiveness of the available information.

The juggernaut of the memory discussion then erupts as to whether the digital archive becomes a humanitarian check-and-balance on democratic governments by curtailing the spread of contemptible information control and corrupt historiography, or does the digital archive rob digital citizens of an essential function of digital forgiveness and societal "forgetting?" It seems European and United States scholars may disagree about these core tenets, yet the search engines of the modern web-based archive do not portray an experiment to prove these notions "right" or "wrong." Policies of web governance shape the way information is archived and how it may be retrieved by individuals for decades to come. Let us ask, "What are the long-term implications of ripping memory holes in the web?" Is this condition more desirable than a constant memory that leaves no room for reputation formation and preservation, forgiveness, or an individual's growth or change through a period of time or a lifetime?

Mayer-Schönberger addresses the individual's loss of power in the digital archive, but then presses further to suggest that digital remembrance negates the convention of time and thus influences people's ability to make rational decisions. If this point of view comes to fruition in the digital environment, a corresponding impact on the marketplace of ideas occurs. An overarching presumption of the marketplace of ideas is that with proper information, humans can act rationally. The impairment of rationality predicted by Mayer-Schönberger (2009) may influence the marketplace of ideas and further erode the integrity by corrupting the capacity of citizens to make these informed decisions. Digital memory can revive events the human mind has forgotten and the natural course of fading memory gains crutches from online search engines. Yet, without the ability to forget, Mayer-Schönberger (2009) argues, we lose the ability to generalize, conceptualize, and act, since the preponderance of external memory inevitably prompts memory of events we thought we had forgotten, and too much of the past comes cascading to the present, shifting our decision-making ability and causing us to distrust our own memory. External digital memory therefore becomes "synthetic," making it susceptible to manipulation (Mayer-Schönberger, 2009). Through this decontextualization, that understanding and perspective becomes lost. From a societal perspective, the impact then becomes that memory impedes change and the use of external memory, once reserved for special circumstances, has become the new norm. In what he calls the final wrinkle regarding time, Mayer-Schönberger predicts that through the prevalence of digital memory, we may suspend belief in our own human memory and thus intellectually elevate digital memory to the primary way to reconstruct the past. He uses a power dynamic to explain how digital memory creates a chasm between the information rich and the information poor and how comprehensive digital memory shifts societal debate.

Most importantly, comprehensive digital remembering collapses history, impairing our judgment and capacity to act in time. It denies humans the chance to evolve, develop, and learn, leaving us helplessly oscillating between two equally troubling options: a permanent past and an ignorant present (Mayer-Schönberger 2009, p. 127). For European citizens who petition for removal, an unanticipated result can occur—more recognition. When an individual successfully removes data from the search engine after passing through the application process and vetting by the Google "investigator," that person may unknowingly draw more attention to the unwanted episode. Media outlets with de-linked articles may draw attention to the fact that it had fallen to the right to be forgotten, thus underlining the original offense. This phenomenon emerges in the right to be forgotten dialogue and has been called the Streisand effect (Brock, 2016), named after the famous singer who attempted to suppress photographs of her Malibu home. The very nature of attempts to suppress propel awareness to a much higher level. In the case of the right to be forgotten, Spain's Costeja Gonzalez, the original plaintiff who launched the right to be forgotten with his 2010 lawsuit, suffers the ultimate Streisand effect.

The clamor for access to the digital archive and the robust web memory resides in a general need to understand the other. Societally, we are seeking to know *who* is that person we work with? How can I learn more about that person's habits, personal life, hobbies? What may we have in common that could bind us together? We have a longing to engage with the other in the present sense. While we are all citizens, some have achieved a new social and legal status—that of a public figure with profound name recognition to a public official elected by the people to serve the public. Some scholars have attempted to decipher this classification as "public" or private individuals, delineating their ability to request removal of information. In a laudable attempt to apply the body of United States libel law to the right to be forgotten, as mentioned in chapter 4, Rustad and Kulevska (2015) suggested a potential framework. For Wechsler (2015), the harm to others must be considered in the data-removal process, most notably, the information seeking public and the original content distributor whose data was removed. In the absence of an appeal mechanism for Google takedowns, the scholar suggests the European public may make a claim under Article 10 of the European Convention of Human Rights. This article guarantees the right of Europeans to receive and impart information. The article states, "Everyone has the right to freedom of expression. This right shall include freedom to hold opinions, and to receive and impart information and ideas without interference by public authority and regardless of frontiers" (European Convention on Human Rights, 2017).

Some may thus argue that the right to be forgotten patently infringes on these rights (Wechsler, 2015). While the text of the ECHR states individuals should have the right to receive and impart information without interference from a public authority, the question then becomes one of state action. In a two-prong response to the classification of Google as a data controller, Wechsler (2015) asserts first that the member states should prohibit private actors from violating the rights and second, the states should maintain responsibility for relegating the authority to Google, a private organization now functioning as a public authority. While scholars and business executives attempt to devise interim and long-term solutions, they face daunting challenges.

The advent of the World Wide Web has put the Information Society on hyper-drive. With a few keystrokes, we gain access to a wealth of information, giving us an enlightened viewpoint on any topic imaginable. From the moment we complete all characters for the word by snapping it into the search engine bar, we expose ourselves to information that is already algorithmically biased, based on our search engine history, with economic forces suggesting products of interest to us, and now today for Europeans, the information sought may have already been de-linked, creating a memory hole in the availability of information. In an examination of the impact of the right to be forgotten on history and collective memory, we find troubling philosophical and social dilemmas. The distortion of history though the search engine de-linking recollects the Orwellian notion of "memory holes," when the Ministry of Truth adapted reality to fit the tenets of the totalitarian regime. We can take a lesson from this totalitarian movement and heed a warning that distorting and re-writing history may take democracies down a slippery slope of totalitarian activities, as philosopher Hannah Arendt has warned. By faking the history of a nation, totalitarian regimes can align the society's memory to fit the totalitarian propaganda. The new digital archive denotes a societal shift from a state of forgetting to one of memory. The right to be forgotten provides a legal framework to give European citizens a recourse to erase past information. While some view this as a noble gesture, the virtual act of de-linking robs other individuals of the right to access and erases the "view-ability" of the information.

Yet, the imperative to examine the right to be forgotten in the context of Big Data remains, as new evidence reveals new adaptive models of decision-making for economic, social, and political purposes. The global struggles for data governance typify the growing pains of the global information society. For U.S. citizens, the First Amendment provides a compelling, immoveable shield against censorship and haphazard efforts to remove data from online search engines. However, American scholars and jurists cannot enforce what is happening in Europe, because Google and other search engines honor the regulations that allow for search engine de-linking and practical removal of information from the World Wide Web. We can only cajole advocates for the free flow of information to view the right to be forgotten not as a tool of personal privacy, but rather as a threat to information freedoms that we have come to expect in the Digital Age. We no longer use the data in a discretionary manner, but we now survive on daily consumption of news and information. Does the right to be forgotten thus threaten a human right to information access? Will we allow European hegemony to shape the availability of information? These are questions that we, as advocates of the free flow of information, must ask ourselves. To further understand the generalized argument that the right to be forgotten addresses a privacy issue, the next chapter explores the intricacies of the privacy argument.

Chapter 6

Refuting the Privacy Argument

Every day, citizens willingly upload information about every facet of their personal lives to the World Wide Web, telling stories about arguments with loved ones, personal vacations, achievements, and sorrows. Through social media outlets, friends become an "audience" to share these intimate tidbits and we peer into the lives of others by examining news feeds, photos, and tweets from our universe of friends. At first glance, we may assume we have raised the barrier of tolerance for the public display of private information. We document all major happenings in our lives with still images and videos, sharing them across multiple social media outlets with pithy quotes about our lives. But occasionally, a friend may post a picture that doesn't gel with our personal reputation or brand. A late night photograph posted to Instagram or Facebook seems like a good idea at the time, but morning's light reveals that it doesn't put your best foot forward for that future employer or relationship partner. The threshold collapses quickly, when we contemplate the ability to take information down that is embarrassing, or no longer true, and we wish we could omit the information from the digital history of our lives.

The implementation of the European General Data Protection Regulation has prompted government officials, privacy activists, and scholars to use privacy law as an instrument of enforcement for data protection. Today, debate centers not only on the legal implications of the right be forgotten, but also the cross-cultural differences between the United States and Europe, when considering notions of privacy and the corresponding legal challenges that emerge when privacy rights collide with the First Amendment and freedom of information. U.S. legal scholars and attorneys struggle to understand how our privacy law may apply to parallel notions of the European right to be forgotten. Is it appropriate to invoke the notions of privacy into the broader concern for search engine de-linking? This chapter defines the tension and points to the conflation of privacy arguments with the broader substantive issue of freedom of information in society. Emerging from this chapter, the reader may better understand the merits of a counter perspective on the application of privacy law to the right to be forgotten circumstance by examining the multicultural views on society and the impact on privacy, and exploring the body of scholarship that attempts to apply privacy law to search engines such as Google, Bing, and others.

The vocabulary between Europeans and Americans about privacy demonstrates that the privacy discussion differs vastly between citizens on opposite sides of the Atlantic. We utilize different privacy lexicons. In what has been called a different understanding between personal liberty and personal dignity (Whitman, 2004), Americans gravitate to words such as "liberty," "reasonable expectation of privacy," and "protection against government intrusion," when contemplating the practical nature of privacy. In contrast, Europeans and laws in the European Union focus on an individual's "dignity" and "reputation." We speak two different languages and possess vast divergences in our points of view regarding privacy. These varying viewpoints contribute to a dichotomy between European courts and United States courts, when balancing freedom of speech with privacy concerns. The cultural values and historic legal heritage has led European Courts to focus on the dignity rights of the citizen more so than free speech rights, thus creating a fertile ground for the creation of the European General Data Protection Regulation and the right to be forgotten. Today, the two Western cultures seem on "irreconcilable" paths toward the right to be forgotten, with Werro (2009) calling it a "transatlantic" clash. Recollecting the origins of American and European jurisprudence provides additional insights into the opposing viewpoints and accompanying cultural traditions. The law of the United States developed with a skepticism for centralized power, while European law was influenced by the French notion of aristocratic honor and German ideals of autonomy and self-determination (Whitman, 2004). This American distrust takes shape in the construction of the Fourth Amendment, which prohibits illegal search and seizure at one's home and further underscores the rights of the property owner to experience a safe haven inside the walls of one's home. Yet, vast ideological differences between the United States and Europeans may be summarized this way: "One way to look at this divide is in terms of the cultural loci of trust: The Europeans trust in the government and distrust the market. Americans take the precisely opposite view" (Werro, 2009, p. 299). Put succinctly, personal data in the United States resides in a contract/property paradigm, while Europeans view personal data in a "rights" paradigm, creating a "cultural gulf" in privacy law (Peltz-Steele, 2015), one that will not likely be bridged by a data protection agreement. The debate quickly moves from an academic or legal one to a discussion of morals and the ability for individuals to seek some type of digital redemption. For example, it has been asserted the ability of individuals to take down information from search engines is a reclaiming of power over one's image, particularly for young people who publish too much information, making mistakes at an early age (Moore, 2017). Individuals exhibit broad levels of tolerance for personal information placed online.

This text assumes the North American perspective that advocates for the free flow of information in society and dissects arguments to explain why the right to be forgotten offends the First Amendment. For Europeans, this book presents an alternative view on privacy and the free flow of information. For United States citizens, this book emphasizes the need to preserve the free flow of information in the face of external intrusion to gain light and galvanize further activism. By learning of this threat, United States citizens may turn attention to activism and advocacy. Despite the uneasiness shared by American scholars about the infringements on constitutional provisos, discussion persists on how to evolve contemporary understanding of privacy laws in America. The two concepts of privacy have become convoluted in the discussion of the right to be forgotten, because different global audiences arrive at the meeting point with culturally relevant viewpoints, historic traditions, and legal heritages. The privacy protections for the United States citizens, not clearly defined in the U.S. Constitution, have emerged from case law and practical jurisprudence administered by the court system, which has led to what can be called a "conundrum" or "paradox" in the vernacular of privacy protections in light of digital redemption, erasure, or oblivion.

PROTECTIONS FOR PRIVACY IN THE UNITED STATES

A United States citizen may state claims to a friend or family member, "I have a right to privacy," and, "I have a right to speak my mind." Citizens may not recognize that one of these rights is specified and legally protected by the United States Constitution, while the other is not. Freedom of speech is a recognized right in the First Amendment of the Bill of Rights of the Constitution, but the word "privacy" does not appear in the Constitution. The concept of privacy has been applied to and derived by legal scholars from incorporation with the Fourth Amendment, protecting the sanctuary of the home from illegal searches and seizures (U.S. Constitution, Amendment IV):

> The right of the people to be secure in their persons, houses, papers, and effects, against unreasonable searches and seizures, shall not be violated, and no Warrants shall issue, but upon probable cause, supported by Oath or affirmation,

> and particularly describing the place to be searched, and the persons or things to be seized.

How do we derive this right in the United States? The answer to that question presents a complicated walk through case law that typically begins with a discussion of a 1890 *Harvard Law Review* article (Warren & Brandeis, 1890), wherein the article describes privacy as a "right to be let alone." While they were addressing the disruptive technology of Kodak cameras, Warren and Brandeis carved out a space of privacy for the individual, drawing from other doctrines associated with slander, libel, trade secrets, and intellectual property (Blackman, 2009). Privacy law was expounded into four torts, which were solidified in 1960 by William Prosser, a former dean of the University of California at Berkeley law school. Now known as the "Prosserian torts," they include appropriation, intrusion, false light, and public disclosure of true, but embarrassing facts (Prosser, 1960). Today, most states recognize at least one of the four distinct privacy actions defined by Prosser (Siegel, 2014). While the old adage that beauty is in the eye of the beholder, privacy resides in the eye of the beholden. Americans reside in a landscape punctuated by various concepts and privacy case law, looking to individual communities to define "what is" and "what is not" privacy.

Americans point to the Fourth Amendment text of the U.S. Constitution to depict the elevation of the home to a bastion of privacy that cannot be penetrated for illegal search and seizure. The notion of the removal of interference to make way for a peaceful state of being "left alone" has prompted other scholars to suggest some precedence may be found in American legal heritage and subsequent case law to suggest the feasibility of a right to be forgotten to emerge in the United States in some fashion or another (Stenning, 2016). The tiny roots of "forgetting" in case law may be found in a 1931 California Appeals Court that found in favor of a former prostitute seeking relief from a movie producer who dramatized her life circumstances that she had moved beyond (*Melvin v. Reid*, 1931). American case law allows for broad coverage of working news media, protecting them from government intrusion in reporting. The exercise of this traditional freedom holds the door open for the marketplace of ideas. There were two key legal cases in the United States that further bolstered the rights of the media at the expense of potential privacy rights of individuals. These cases, *Florida Star v. B.J.F.* (1989) and *Cox Broadcasting v. Cohn* (2017) established a significant framework to protect the reporting of information that has been entered into a public record.

In contrast, the Charter of Fundamental Rights of the European Union grants to European citizens the right to privacy in Article 7, as stated: "Everyone has the right to respect his or her private and family life, home and communications." The right to privacy also extends into data protection in the Charter's Article 8 provision as stated: "Everyone has the right to protection of personal data concerning him or her." The appropriateness of treating data protection as a "fundamental right" remains under dispute (Van der Sloot, 2017). While human rights provisions may garner "absolute" status, data-protection rules are more malleable, permitting state regulation at times. By revealing the broadening of regulatory power over data protection in Europe and pointing to the phenomenon that data protection seems to protect "something quite different" than traditional human rights, Van der Sloot asserts that Europe's General Data Protection Regulation results in market regulation, not the protection of human rights in Europe. Following this logic trajectory, data protection may default to an auxiliary position as a typical consumer right, not a fundamental one that garners additional legal protections.

PRIVACY PERCEPTIONS IN EUROPE AND THE UNITED STATES

The tug of war between privacy activists and First Amendment advocates continues today. The tension intensifies in the right to be forgotten landscape. A conflict of privacy erupted in the United States when members of the Westboro Baptist Church protested the funeral of Marine Lance Corporal Matthew Snyder. Standing on a public street, Westboro members carried signs stating: "America is doomed," "Fag Troops," and "Semper Fi Fags" during the funeral. Appalled, Snyder's family asserted in the court that the context of the speech, in this case the funeral, was not appropriate for the language. While a lower court awarded Snyder's family damages for emotional distress, the U.S. Supreme Court affirmed the reversal of the district court, finding that the despite the repugnant nature of picketing at a funeral, the protestors remain protected from liability by the express protections of the First Amendment. In its decision, the court noted that some "purely private matters," such as an individual's credit report, do not merit First Amendment Protection (*Snyder v. Phelps* 562 US 433, 2011). This finding suggests a hint of forgiveness and a zone of privacy protection for individuals. The Westboro Baptist Church victory affirms the longstanding belief that citizens have a right to express themselves in a civic environment, particularly when the matters were of public record, were not provably false, and were expressed through hyperbolic rhetoric. Writing for the Court's majority opinion, Chief Justice John G. Roberts, Jr. wrote, "While these messages may fall short of refined social or political commentary, the issues they highlight—the political, and moral conduct of the United States and its citizens, the fate of our nation, homosexuality in the military, and scandals involving the Catholic clergy—are matters of public import" (*Snyder v. Phelps* 562 U.S. 433, 2011).

While we ponder matters of political concern as protected speech, let's advance to another Supreme Court case that has long-lasting implications for the practice of political speech, the First Amendment, and perhaps the administration of the right to be forgotten in the United States. *Citizens United v. Federal Election Commission* (558 U.S. 310, 2010) found that, while the government can regulate corporate speech through disclosure requirements, the government cannot limit the speech altogether. The case dealt with the dissemination of the film, "The Movie," providing content on then Senator Hillary Clinton and her fitness for political office. Citizens United sought an injunction against the Federal Election Commission to prevent the Bipartisan Campaign Reform Act (BCRA) from excluding corporations from funding such political communication as election advertisements. The Supreme Court, in a 5–4 vote, found that under the First Amendment, the corporate funding of independent political broadcasts cannot be limited. This action opened the floodgates for corporations to serve as "communicative citizens." This case personified corporations as citizens who can articulate "speech" in the contemporary marketplace, particularly political speech. The Citizens United case casts a shadow onto the protection of search engine results, as we consider the algorithmic speech generated by Google, and may be appropriately classified a corporate speaker with First Amendment protections.

While constitutional questions in the United States erupt in light of the right to be forgotten, further exploration of the privacy argument reveals how this line of argumentation presents a smokescreen for the larger issue of information availability. The right to be forgotten has set the agenda for a robust discussion of privacy rights inside and outside of Europe (Youm & Park, 2016). Europeans embrace the blanket of privacy, and in the United States some privacy activists have suggested that the failure of the United States to embrace a right to be forgotten represents an unfair business practice (Consumer Watchdog, 2015). The unfair business practice erupts over concerns that U.S. users of the Google search engine are not treated in the same way as the European counterparts. While European audiences may seem perplexed about the unwillingness in the United States to embrace privacy practices that in Europe are deemed reasonable and necessary, the origin of the debate resides in conflicting interpretations of privacy. The divide cuts across both sides, as Europeans contend that U.S. citizens do not understand the imperatives of privacy, based on a lack of privacy etiquette (Whitman, 2004). The dichotomy of the European notions of "dignity, preservation of honor, and personal respect," contrasts with the U.S. notions of liberty and protection against government intervention (Whitman, 2004), creating a rift that legal governance may not successfully traverse. Further, scholars suggest that countries place more legal weight behind free speech in contrast to privacy rights in Western democracies correlate with whether the society's legal system has emerged from a common law or civil law tradition (Kelly & Satola, 2017), and, while the United States has emerged from a common law heritage, the multitude of European countries are derived from a civil law tradition.

As we navigate the intersection of the right to be forgotten and the First Amendment, prompted by the European right to be forgotten, it is relevant to examine two historic media law cases, both of which involve the naming of a rape victim. The first case, *Cox Broadcasting v. Cohn*, was decided by the Supreme Court in 1975, and was pursued by a broken-hearted father, Martin Cohn. His 17-year-old daughter had been raped and killed, and reporters working for Cox Broadcasting Corporation obtained her name from a public record and used it in news broadcasts. The father purported that the case violated a Georgia privacy statute, which prevented the use of rape victim names. While one can certainly appreciate the father's difficult position, the court found that, indeed, the Georgia statute violated the Constitution's First Amendment, and it was determined that any privacy issues "fade," when information appears on a public record (*Cox Broadcasting v. Cohn*, 1975). In the second case decided by the U.S. Supreme Court in 1989, *Florida Star v. B.J.F.*, a reporter printed the name of a sexual assault victim despite a Florida law prohibiting such action. While the district court in Florida found the newspaper guilty and awarded damages, and a district court affirmed the decision, the U.S. Supreme Court reversed the decision. The court found the damages violated the First Amendment because the information was lawfully obtained and truthful. Further, the Florida statue itself was deemed invalid because it did not protect a state interest. These two cases fall into the bullseye of privacy, because sexual assault victims seek anonymity so the victims and their families do not have to constantly live the crime over again. However, the use of the public documents in news-gathering provided the constitutional coverage that was needed to make these statutes fall. The precedence established in these cases and others provide a central rallying point because the United States now considers the right to be forgotten as conflicting with privacy interests.

The climate after the Florida Star case suggests it is very difficult, if not impossible, for states to regulate news media from publishing information as long as it was legally obtained. The *Florida Star* case suggests that, while media outlets are legally allowed to publish sensitive information, ethical codes may regulate such behaviors. Today, ethics codes guide professional behavior and provide guidelines for individuals working for reputable organizations in journalism and public relations. Disclosure of information is a core principle found in the Public Relations Society of America member

Code of Ethics (PRSA, 2017). The code states, "open communications fosters informed decision-making in a democratic society. The intent of the core principle is to build trust with the public by revealing all information needed for responsible decision-making" (PRSA, 2017). The example of improper conduct for this provision is "lying by omission." For example, the code states that a public relations professional who knowingly fails to disclose required financial information, painting a misleading picture, would violate the code (PRSA, 2017). Further, the International Association of Business Communicators also set forth a code of ethics that recommends members engage in consistent, responsible, ethical, and legal choices in all of their communications (IABC, 2017). For some, the right to be forgotten may be considered a European legal force that compromises truthful and accurate communication by concealing information and removing available data from the marketplace. The Society of Professional Journalists (SPJ) code delineates that journalists have a special obligation to serve as "watchdogs" over public affairs and that journalists must "seek to ensure that the public's business is conducted in the open, and that public records are open to all" (SPJ, 2017). In addition to serving as watchdogs for the public, the code asks working media to "balance the public's need for information against potential harm or discomfort," and show compassion for those affected by news coverage (SPJ, 2017). Instead of legal restraints, the United States relies on professional ethics to moderate activities and preserve integrity.

Certainly legal solutions for privacy are not the only options, because society still maintains social rules and mores about human decency and etiquette. Even adherence to ethical codes encounters challenges, however, when faced with an external legal mandate like the right to be forgotten. Scholarly analysis of common codes governing journalists and information professionals suggest that the right to be forgotten violates these ethics codes by suppressing freedom of speech and imperiling the marketplace of ideas (Byrum, 2017). To reconcile the various notions of privacy between Europeans and Americans, Walker (2012) suggests we consider the debate as a spectrum between dignity and liberty and that discussions should more aptly focus on *who* will protect the privacy. In Europe privacy is administered through state intervention, while in the United States an individual citizen is more likely to act independently and advocate on her or his own behalf. The ideological rift persists across geographic borders and should prompt citizens on both sides of the ocean to contemplate the actual impact of implementing the right to be forgotten.

SITUATING THE RIGHT TO BE FORGOTTEN IN A PRIVACY CONTEXT

The right to be forgotten solidifies the notion of "practical obscurity" under the law, because individuals exercise a sense of dignity by determining the content and timing of information sharing (Youm & Park, 2016). With these disparate views, the tugging occurs, while Google and other search engines remain legally bound to implement the European General Data Protection Regulation by honoring requests for information de-linking. In Europe, the wingspan of privacy now casts a shadow over the data regulation, yet in reality, one must question if the application of the right to be forgotten is truly a privacy concern, or a dire threat to freedom of information. Instead, as search engines cross international borders and the eradication of certain pieces of information create an inaccurate portrayal of data retrieval in a global information environment, the keen attention to digital privacy promotes emerging troubles for the free flow of information in society today. It has been asserted the right to be forgotten may compromise the integrity of Internet data through the creation of "memory holes" (Antani, 2015), and chapter 5 addresses this dilemma in greater detail. Compounding these issues, the situation further empowers the data controllers, inculcating them with significant power over preservation and retrieval of data. The current enforcement approach to the right to be forgotten allows for private companies to determine if de-linking will occur, shifting power and accountability away from the national data protection agencies (Wechsler, 2015) to private enterprise.

As we discern the implications of the right to be forgotten's impact on data retention, analysis, and storage, search engines by default inherit vast amounts of influence and authority in the EU, because they become the decision-makers on the fate of data. In practicality, the right to be forgotten could lead to corporate censorship, because Google may seek to minimize fines or legal actions and cautiously remove ambiguous cases (Simpson & Bates, 2015). In what have been coined as "deciders" (Rosen, 2012), the individuals at Google and other search engines wield a magnified and authoritative power over information management.

In an interview with *New Yorker* magazine writer Jeffrey Toobin, Viktor Mayer-Schönberger (2009), cites the European resistance to data collection and retention as a reaction to historical events, including the German totalitarian "surveillance society" that placed power into East Germany's Stasi, which captured, catalogued, and used citizen data in their enforcement actions. After the fall of the Berlin Wall in 1989, emerging democracies wrote laws to prevent data collection and retention abuses by government agencies and strengthened privacy laws (Toobin, 2014). In his own book, Mayer-Schönberger asserts the privacy concerns become more acute in the Internet Age, pointing to the fact that, while the Stasi kept records in hard copies and film, the contemporary era composed of digital files, online storage, and data retention in the cloud make information more durable and readily retrievable. The adherence to stricter data retention and preservation protocols has certainly found its way into the European General Data Protection Act.

New Audience, New Permanence, and De-Linking Requests

Acceptance of web-based publishing methods has changed the way we convey our personality. We readily upload personal information and preferences about our favorite restaurants, fashions, sports teams, and an unending list of personal interests and attributes. The audience for our universe of user-generated content is now much larger than our offline social group of friends—now we can tell virtual "friends" all across the nation about our recent published article or rant about a new political policy. The audience expands exponentially, while the mediated environment establishes a membrane between the sender and the receiver. Citizens today use mobile devices to publish rich information about themselves and others, sharing the data across vast distances. The mediated environment often removes a veil of the tactfulness found in traditional person-to-person, and online conversations can rapidly decline to name-calling and proliferation of false or misleading information. Therefore, the nature of traditionally private thoughts changes when we willingly submit information and place ourselves into the vortex of information that swirls around online search engines. The absence of control over personal information becomes a sticking point for those seeking greater privacy provisions. The permanence of online content changes the game, unlike the fleeting nature of the static newspaper or local news show that was quite transient and easily fell away from our memories. The average citizen could not formerly gain access to the newspaper's archive or the television station's library of past broadcasts, unlike today, when such historical data is readily available for the individual with a computer or smart phone. The availability of content has mushroomed, as technology tools such as YouTube and the Facebook timeline make creating and sharing your analog world easily transfer to the digital world with a few clicks. The right to be forgotten may also gain consideration as a tool or recourse for surveillance. Therefore, for the sake of argument, we can contemplate the web search engine as a powerful data surveillance tool for citizens, leading to vast implications inside the borders of the United States and beyond. For individuals who have committed crimes, Europe offers more "forgiving" regulations, while the web provides a perpetual memory of past criminal behavior in the United States.

Early Removal Requests

In 2009, after successfully getting the article removed from the German language version of Wikipedia, two German killers sued the Wikimedia Foundation, Wikipedia's parent company, in an attempt to get the global medium to take down the English language version of an article about their trial and conviction (Schwartz, 2009). In line with German legal customs and precedence that allow convicted criminals to enjoy privacy after serving their sentence, they sought an expansion of complete removal beyond Germany's borders. The case became an "instant online cause celebre" (Arthur, 2009), because German privacy laws mandate that after a certain period, a crime is "spent" and cannot be cited, allowing the criminal to enjoy an aspect of privacy and "the right to be left alone." This case represents the finer nuances of contemplating privacy across global borders. While German citizens embrace a notion of forgiveness, United States citizens do not readily compromise First Amendment rights for privacy concerns. Meeting a maelstrom of First Amendment protest, the pair did not prevail, and today the Wikipedia site accessed from the United States continues to include a story about the involvement of Wolfgang Werte and Manfred Lauber in the murder of actor Walter Sedlmayr, including updates on the legal machinations that ultimately failed to get the material taken down in the United States (Wikipedia, 2017). In the United States today, digital news and information evolves into a state of permanence, and the issue of digital surveillance continues to agitate and alarm individuals, further complicating perceptions of the right to be forgotten.

Omniveillance and Surveillance

For privacy activists, digital surveillance may seem like it is coming from everywhere. In recent years, Google's Street View technology has drawn criticism as an over-reaching invasion into our private lives, as Google vehicles drive down city streets, photographing the exterior of homes, streets, public places, parks, and other publicly visible landmarks. Under the current United States law, the safest place for a journalist (or whomever) to collect and report information is on a public street, just as Google employs the public street to gather the information. The marked difference today is the 24/7 availability of the information via the web. Blackman (2009) attempts to define "omniveillance" as omnipresent digital surveillance in public places, thus diminishing an individual's solace or newsworthiness. In an effort to draw an analogy between a public street and the web, the argument collapses. The public street is just that—public. The online search engines used to retrieve data about an individual are privately owned. While they are considered parts of public companies that sell stock to the public via the NASDAQ stock exchange, these companies are not state-managed utilities. Privacy activists still try to construct an argument against omniveillance by asserting that the constant surveillance, in fact, suppresses freedom of expression. In public, individuals may experience what is known as a "cloak of anonymity" or a reasonable expectation of privacy, which allows them to conduct themselves with greater freedom of expression. When someone is watched, however, they may suffer a loss of that unbridled sense of freedom and self-censor themselves. This argument leads to Blackman's (2009) suggestion that in a world devoid of privacy, where "everything can be known to everyone," free expression suffers because the individual will withhold speech for fear of embarrassment or ridicule, casting a chilling pall over expression. Therefore, following this line of argument, a modicum of privacy may be required to protect freedom of speech.

Let's view this argument through the lens of the European right to be forgotten to determine lessons about how this may harmonize (or not) with the freedom of expression in the United States. The argument goes something like this: The web search engine constantly surveys all events and actions you have placed into a public or quasi-public online domain, allowing the engine to readily recall information in just a few seconds. This dominance presents numerous social implications, most notably the willingness to preserve "face" by donning a perfect mask that only reports the most positive news about our lives and delimits the negative aspects. But, negative information eventually will weave its way through the data bytes and find itself regurgitated online. Privacy activists clamor that an individual should have a right to take down this data, perhaps because it became outdated, is no longer relevant, or is published by another person. Has the omnipresent gaze of the web really impinged upon our freedom of expression or, ignoring possible impingement, do we continue randomly upload messages, photographs, and images to the web to allow our audience to view? This may be where the argument that surveillance promotes self-censorship, which can lead to impairments of freedom of expression declines in the era of frenzied social networking. Today, citizens willingly place more information into the public via social media. The fact that Google Street View and the Google search engine represent the same enterprise, the power over control of data seems staunchly placed into the hands of Google. While scholars have suggested that the "Orwellian future of omniveillance can lead to the evisceration of privacy in public" (Blackman, 2009), a web observer can only wonder about the implications of Google managing Street View and simultaneously having the power to determine what stays or goes from the online search engine results. This may morph into a monopolistic exertion of information control that is not government based but rather government sanctioned by the right to be forgotten. The discussion of omniveillance and surveillance ushers a new question: Are search engines a form of societal surveillance? As Digital Age citizens, do we willingly and purposefully submit information online to subject ourselves to higher levels of surveillance? Perhaps . . . but, if not, it is right there in the Google user agreement. In line with Nissenbaum's assertion that we need to consider the context of the information, data submitted by the user and placed onto a social media site is legally covered by the user agreement. Citing availability of online pubic records, consumer profiling though data mining, and the prevalence of radio frequency identification, Nissenbaum (2004) points to technology tools that chip away at previously conceived notions of privacy. The purchase of home robots may also pose future privacy threats, as they mature as data collection devices (Kaminski, Rueben, Smart, & Grimm, 2017).

The market dominance of Google creates an environment whereby control over information becomes centralized. Grimmelmann (2008) has succinctly stated: "Whoever controls the search engines, perhaps, controls the Internet itself. Today, no one comes closer to controlling search than Google." In what he calls the "Google Dilemma," Grimmelmann points to the fact that Google has authority to "shape" search engine results, posing both legal and ethical challenges. Citing a case involving Turner Broadcasting System, Bracha and Pasquale (2008) recall the First Amendment does not disable the government from taking steps to ensure private interests don't restrict the free flow of information. In contemplation of the "must carry" provision of the 1992 Cable Television Consumer Protection and Competition Act that required broadcasters to devote some of their channels to local broadcasts, the court found that Congress has an "independent interest in preserving a multiplicity of broadcasters" thus supports Congress' right to determine facets of a competitive marketplace (*Turner Broadcasting Inc. v. FCC*, 1997). The argument for regulation does not necessarily translate from cable television into the web (Siegel, 2014), because the web offers an infinite number of information sources, operates with a relative lack of online gatekeepers, and provides a low cost of information production. Despite these differences, the evolution of the web as a means for transmission of mass communications messages requires innovation in communication law, including new regulations for online libel and intellectual property law, which protects trademarks and copyrights online (Siegel, 2014). Legal analysts and practitioners must consider the emerging technologies and decipher new approaches that are in harmony with the United States Constitution.

Digital Footprints and Big Data

As we submit to a multitude of online transactions, whether posting information on social media or emailing a colleague, we leave digital footprints with traces of our information residing on servers humming in a chilled room. Large servers often have redundancy, which replicates our digital traces overseas and keeps data constantly churning through the digital atmosphere. Even if we long to be forgotten and place our data into oblivion, it will not really happen—because data removed may find its way to a database for de-linked information. This data never really goes away, except through the force of digital decay—broken links of a website no longer hosted, databases corrupted by a multitude of factors, including everything from a flood to the server farm, or the leftovers from a predatory hack. As digital citizens, we leave breadcrumbs about our interests, our habits, our net worth, our spending thresholds, and more. Through the use of this data, predictive analytics foretell a "Selling Machine" of hyper-suggestibility, as we take a look at a vacuum or a chandelier on eBay and the next day, and the following day, Google serves up an advertisement for the item. Savvy marketers say this marketing technique enhances the user experience, makes our lives easier, and greases the wheels of American consumerism. At the same time, privacy advocates condemn the practice and urge consumers to disable cookies and short-circuit efforts to fully exploit the digital trace. Considering this trajectory of online data, are we poised to have an online right to be forgotten for our purchase history? Or, can we erase our online surfing habits to give us a "clean slate"? This is a privacy concern that citizens advocate in the online realm. The logic for this is predictable. That is why the user agreements of online commerce and social networking sites attempt to disclose the intended use of Big Data before you go clicking around . . . and before you have the notion that you may want to ultimately remove the information at a later date. To counteract these infringements, Jones (2016) points to "obfuscation" as a way online technology users can maintain what she calls "contextual integrity." Common obfuscation techniques may include using a pseudonym, exercising strict privacy settings, encrypting passwords, deleting cookies, and utilizing multiple accounts. Moving beyond obfuscation as a way to hide, Jones calls the right to be forgotten a technique of "information obstruction" and points to the resulting effect that the right to be forgotten impairs the discoverability about one's past. While Jones call this an "advantage" because obstruction supports the creation of a dynamic self and maintains contextual integrity (Jones, 2016, p. 94), First Amendment activists find obstruction to information flow contradictory to existing laws. The ability to "obstruct" the availability of information in the amorphous digital archive retrievable by the common search engine flies in the face of society's free flow of information and can only fall under First Amendment scrutiny. World famous hacker (now reformed) and the U.S. government's most wanted hacker, Kevin Mitnick, elaborates on Jones's notions of obfuscation in his book, *The Art of Invisibility* (2017). Mitnick advises readers to encrypt email, protect data with password management, hide the IP address from geographical places visited, obscure computers from tracking, and defend personal anonymity while online.

Understanding the Application of Existing Privacy Laws

While contemporary jurisprudence refers back to the four torts of privacy (i.e., appropriation, intrusion, false light, and public disclosure), the movements toward censorship in the online environment provide a continual beating drum. In his own personal blog, Google's attorney, Peter Fleischer, cites the prevalence of privacy contentions as the newly popularized arsenal for privacy advocates, describing it as the "new black for censorship fashions" (Fleischer, 2011). Google remains in a tough position navigating both the European and United States markets. This may ultimately influence their business model. The burden of legal regulation and content liability may impair Google's ability to innovate as they remain in an uncertain, changing atmosphere (Hurst, 2015). Some take an opposing view by attempting to harmonize privacy with existing legal frameworks. Walker (2012) asserts data privacy should be considered in the context of contractual rights in order to avoid First Amendment complications and suggests deleting data that a user has personally submitted may be an act compatible with the U.S. Constitution. He issues a call to United States legislators to create a law that would regulate data protection, particularly data removal requests, to yield supposed benefits. Through statutory creation, legislators would provide clear guidance to all involved, most notably the data operators and the consumers about data usage and protocols for honoring user requests for information deletions. Further, a law would also define legal procedures for enforcement of consumer requests. These types of movements for statutory construction may encounter rapid defeat, when considered in the context of the First Amendment's provision that "Congress shall make no law."

In an effort to make sense of the prevalence of surveillance in public spheres and the corresponding application to privacy, Nissenbaum (2004) asserts existing models for privacy are simply inadequate and unsatisfactory for the Digital Age. Asserting public surveillance violates a right to privacy, Nissenbaum sets forth a new model of "contextual integrity," which links privacy protections with an appropriate context, defining it as "compatibility with presiding norms of information appropriateness and distribution" (Nissenbaum, 2004, p. 155). In this model, she recommends information privacy must examine a multitude of factors, including the situation (context), the nature of the information to the context, the roles of the agents receiving the information, the relationship to information subjects, and under what terms the information is shared. All of this must be considered under the terms of the use of the information dissemination. This proposed model takes into consideration the who, what, when, where, and why of the information that is shared, and this prototype has legitimacy for future discussion for data retention, storage, and perhaps even data regulation and governance in the United States.

RECONCILING PRIVACY AND FREE SPEECH

The ability to reconcile what citizens perceive as online privacy concerns and the constitutional protections of free speech confound legal scholars and digital practitioners. The argument that the right to be forgotten deals with privacy concerns suggests a faulty presumption when viewed through the lens of American law. The items that appear on the web have already appeared in the public eye—they have been published in a news media website or entered into a public document, legally accessible to the public. Based on past legal decisions that journalistic news-gatherers may appropriately use all information, most notably the cases involving naming rape victims, American journalists stand on solid footing when gathering data and reporting the day's happenings. For Google, the advent of social media has created a new wrinkle, with Google's own transparency report indicating that Facebook was the website most often requested for search engine de-linking. While an individual may have willingly agreed to Facebook's terms of use and uploaded information, the person asks for power and control over information created about them, some of which may be unflattering or was posted by another person. This creates a more complicated situation for Google under the right to be forgotten, because the company now becomes the "grand arbiter." But is that really a privacy issue or one of freedom of speech and unencumbered access to digital information? While First Amendment activists may want privacy as much as anyone else, they recognize that legal actions to impose restrictions on freedom of information can threaten democracy itself by compromising the marketplace of ideas. Moving forward, it is prudent for right to be forgotten opponents in the United States to untangle the data regulation from privacy. Let's consider a three-pronged argument that privacy is not the issue at stake in this situation: (1) Government intrusion into the free flow of information as found in the right to be forgotten is akin to government intrusion into the sanctity of one's home. (2) Individuals maintain the same recourse for invasion of privacy, including the four Prosserian torts that allow for legal remedies in the event that information is shared improperly. (3) The information already exists in the public domain. The definition of privacy translates into nondisclosure of personal information to third parties. Privacy is not the real issue; the right to be forgotten's power over information is the real issue.

European Privacy Rights Infringing on Information Availability

Since the ruling in the Google Spain case against Mario Costeja Gonzalez, scholarly debate has churned over the tensions between an individual's privacy right and freedom of expression. Corresponding issues, including Internet governance, the role of private data controllers, and the role of human rights have ushered in a new era of dialogue that will not likely diminish, when the General Data Protection Regulation commences enforcement in May 2018. Through enforcement, Europe trumps freedom of information across geographic borders. The issue of national sovereignty has sparked a new debate as the right to be forgotten values the cultural adherence to privacy rights of citizens over the freedom of expression rights of another country. In July 2017, the European Union's top court agreed to hear the dispute between Google and France's privacy regulator to determine how broadly the right to be forgotten could be applied, thus answering the question if the data removal effectuated in one country should apply to other European domains (Kostov & Schechner, 2017). The issue stems from a 2015 incident in which France's privacy regulator ordered Google to apply the right to be forgotten removals to all of its websites, despite the geographical location of where they are accessed. Implementation of this would impair the ability for others rightfully accessing information in their home country like the United States. Google refused to honor the French authority mandate and was fined. The appeal resides with the European Union's High Court. The complicated legal maneuvering in this case deals with a concept known as "extraterritoriality," thus indicating Europe's digital privacy laws would influence the availability of information outside the region. The data protection regulation enacted in 2017 forecasts a worldwide applicability of European Privacy Law. The question then emerges: Are U.S. citizens willing to allow European hegemony, with blind adherence to privacy rights at the cost of global freedom of expression, to prevail over the United States and its constitutional prerogatives? Scholar Robert Larson (2013) says "No," with three counterarguments, stating that the right to be forgotten diminishes the ability for citizens to participate in the marketplace of ideas, the legal action disrupts the natural process of communication and debate, and it misappropriates market dynamics to a legislative function. Sir Tim Berners-Lee, recognized as the inventor of the World Wide Web, has said, "The right to be forgotten seems draconian when it's about removing information that is true. The idea that information which is true can be removed from public record is frightening" (Lauchlan, 2014).

Google's own transparency report states that the company has a team of experts assembled, primarily based in Dublin, Ireland, to review requests. The team members proceed through a four-stage process upon receipt of the de-link web form request: The team assesses if enough information is available to make a decision, determines if the person making the request is a European citizen or resident, verifies if the requestor's name appears in the search engine results, and then makes a determination if the page requested for removal is "inadequate, irrelevant, no longer relevant, or excessive" (Google Transparency Report, 2017). The company further states that the most common material factors involved in decisions to de-list pages include the clear connection to public interest; the inclusion of sensitive information such an individual's health, sexual orientation, political affiliation, or trade union states; content relating to minors, particularly the commission of a petty crime that occurred while a minor; and the spent convictions/exonerations/acquittals for crimes (Google Transparency Report, 2017).

More Thoughts on Privacy Argument

Today, privacy activists and First Amendment advocates struggle to find common ground regarding online governance, digital redemption, and the ability of private individuals to "forget" their past through search engine de-listing efforts. For American citizens though, the information war will not likely go down without a formidable fight. Google's own lawyers and First Amendment gurus have launched efforts to promote a wider awareness of the right to be forgotten and the de facto appointment of Google as the chief information police officer for Europe. To intellectualize the application of the right to be forgotten in the United States, legal scholars have attempted to apply the full body of privacy law in the U.S. to the "right to be forgotten," providing a potential framework for consideration and application of the regulations, if the regulations were to apply in the U.S. (Rustad & Kulevska, 2015). Others have considered the application of United States privacy tort issues, such as publication of public disclosure of private facts when contemplating adoption of the right to be forgotten (McNealy, 2012). In an examination of the newsworthiness exception to the right to be forgotten, McNealy (2012) argues that this provision is cohesive with the protection of newsworthiness found in the U.S. privacy tort's public disclosure of private facts. Specifically, that public information would be protected as long as it was of public concern and was not highly offensive, morbid or recklessly published. While it has been indicated that forgiveness principles are found in the American legal system, particularly related to bankruptcy, credit records, and juvenile criminal records (Blanchette & Johnson, 2002; Bennett, 2012), attempts to restrain the flow of information via online search engines should not appropriately evoke privacy law in the United States. Instead of an emphasis on privacy, constitutional principles solidified by the First Amendment and the ethics of communication encompassing transparency and free flow of information are at stake with the possible implementation of the right to be forgotten. Movements to protect digital privacy, encompassing personal and financial data transferred and stored online, have unearthed key differences between citizens of Europe and the United States about the notion of privacy.

While we live in a global village, connected by a digital culture that generates a social space, arguments persist that the right to be forgotten affords the individual the power to control one's own identity in the virtual world (Carbone, 2015). Some claim U.S. law already makes provisions for the clearing of irrelevant material, as evidenced by the Fair Credit Reporting Act, which deems debt collection, civil lawsuits, and tax liens obsolete after seven years (Simpson & Bates, 2015). It has also been suggested that the right to be forgotten be considered in a similar way to state tort claims for the public disclosure of private facts (Werfel, 2016). Other laws cited as evidence of privacy in the U.S. legal system include the 1974 Privacy Act, which prohibits personal information disclosure by federal agencies without prior consent; the 1996 Health Insurance Portability and Accountability Act; and the 1998 Child Online Privacy Protection Act (Glazer, 2015). Despite attempts to make sense of the right to be forgotten in the United States within a privacy framework, constitutional concerns remain about the ostensible violation of fundamental First Amendment principles articulated in the Bill of Rights to preserve freedom of speech and freedom of the press.

As advocates of the right to be forgotten exert pressure to balance free speech and privacy in the United States, speculation exists about the U.S. approach being narrowly tailored to include regulation of cyberbullying and revenge porn, as the online privacy issues evolve (Antani, 2015). Protectors of the First Amendment stand on well-embraced constitutional theories and a heritage of protecting the marketplace of ideas. By centralizing information control with a search engine/data controller, the marketplace of ideas, as manifested in search engine results, is in danger of becoming filtered or warped, based on the viewpoints of the censor. The information that appears on these search engines has already seen the light of day and appeared in the public domain, making efforts to curtail it, based on the argument that privacy protects disclosure of private information, seem irrelevant (Cheng, 2016). User awareness and software systems designed with privacy in mind provided tentative future approaches (Koien & Oleshchuk, 2013).

Objections to the right to be forgotten continue to rage. Major objections observed at the time of this book's publishing include the privatization of a quasi-judicial decision, the lack of process transparency, and the reinforcement of a quasi-monopolistic hegemony of the search engine on freedom of information (Dulong de Rosnay and Guadamuz, 2016). In what has been called a "fairly dramatic transatlantic schism" (Werro, 2009), the ability for the right to be forgotten will not likely proceed in the United States as it has done in Europe, stating: "There is no reason to think that a right to be forgotten stands any chance of developing under the current Cox and Florida Star regime in American privacy law" (Werro, 2009, p. 298). Whether or not the United States can stem the tide against the European overtures and legal mandates to sanction de-listing remains to be determined, though in the meantime, the health of the idea marketplace in Europe and beyond hangs in the balance. The Google deciders in Ireland can cast the fate of information tidbits that could be used for European democracies to self-govern and discern appropriate social, economic, and political policies.

Chapter 7

The Free Speech Manifesto for the Digital Age

Seven Tenets to Preserve Information Integrity in Contemporary Democracies

The chilling influence of government interference on information retrieval and the ability of private companies to arbitrarily delete data from search engine results foretell trouble for the free world. Historical insights reveal the connection between censorship and totalitarianism. While the European Union moves forward to promote privacy rights of individuals in 28 European countries, Google and other search engines have been thrust into a complicated predicament requiring the corporation to vet requests for de-linking and remove search engine results, based on consumer requests. This regulation puts the search engine in the business of deciding *what information* is available for the world to view. Is the search engine the new world censor to beat all censors? Time flies as data bytes are removed each day from the availability of information deemed "outdated" or no longer relevant. Who is the grand master to determine if these facets hold true, and what criteria do they use to implement thousands of a la carte decisions as they move across the transom of de-linking requests?

This text provides a backdrop to understand why the right to be forgotten finds conflict when considered through the interpretation of media scholars and those who advocate for First Amendment freedoms. As the right to be forgotten traverses the pond from Europe to the United States, an accompanying array of scholarly discussion and legal debate swirl around the issue. While some scholars attempt to discern a legal framework in which the right to be forgotten is even remotely possible in the United States, others stand resolute in opposition to the action that quite obviously pierces holes in the collective memory of the information web.

As an alternative point of view, this book calls for "holding the world open" with information access that allows for the preservation of the information marketplace and a higher level of global transparency that can fuel not only economic collaborations, but also political cross-cultural awareness and personal ties. As United States citizens and legislators examine the presence of European regulations under the right to be forgotten, they must concurrently consider that this represents an arbitrary shift in power and thus resist this regulation. The European right to be forgotten is a legal directive that inculcates power into a private enterprise. The eye of scrutiny must fall on the movement to amplify the power of a private corporation to determine carte blanche what information is available through the Internet. With scant oversight and self-generated transparency reports, the search engines are positioned to usurp vast amounts of power as they determine who, what, when, and where we see information on the World Wide Web.

While advocates argue that the right to be forgotten is simply a privacy tool, representing a recourse for individuals who do not have alternative means to suppress unflattering information that has since become "outdated or is no longer relevant," the removal represents a form of censorship that is undeniable in the Digital Age. Therefore, to preserve the marketplace of ideas, the manifesto calls to suspend censorship in favor of greater transparency and a healthy information marketplace. Those who believe in the merits of freedom of information must actively protest government intervention into information availability to align with the First Amendment protections of the United States Constitution. The right to be forgotten bestows the ability for foreign governments to contravene the First Amendment through a proxy. The search engines serve as the proxy, policing information, and "deciding" what is news. The specter of government censorship and the ability for private corporations to unilaterally decide the availability of information gives them an unbearable monopoly over information and counters the democratic principles found in the heritage of the United States. The framers of the Constitution and the stewards of democracy have preserved the unencumbered free flow of information in United States society for more than 240 years. The advent of the Digital Age creates a new medium for information that allows for instantaneous global communication. Search engines remain a requisite tool of the Digital Age and should not undergo government intervention that requires them to police the World Wide Web of various forms of content.

While corruption, inaccurate information, and the emergence of "fake news" contribute to the cacophony found in the marketplace, the ability for citizens to have access to the full range of information should not undergo denial at the hands of the government. Instead, citizens may address the presence and viral spread of corruption through transparency by shedding light into the situation. Concealed information precludes proper information vetting, and information removal suggests a total obliteration of the ability to evaluate an individual or societal issue. Yet human beings and citizens reject the notion that the online persona can exist in perpetuity. This notion of evolving into a society that "constantly remembers" represents a transitional shift away from a society that "forgets." The Digital Age demands this societal shift, as memory emerges as a digital archive. For many, this is a double edged sword, giving us the power to retrieve vast amounts of information with the click of a few keystrokes, yet when it comes to our own personal information, we may wish that "forgetting" is possible. The sword maintains these two edges, and in the United States, we have embraced the custom of utilizing search engines and treating them as the "authoritative source" for online information, allowing the engine itself to rank the results and shape our perception of events. When citizens shift thinking and recognize that online information is wrong, no longer available, or contains holes, they are left wanting for accurate information on social, political, and economic issues. The removal of information perforates the availability of accurate information on the Internet, and also significantly erodes public confidence in the just inventory of information.

To ensure the robust exchange of ideas on new media channels, citizens must quell government intervention into freedom of information. Through regulation and governance, individual governments, nations, and regions exert meaningful control over information, calling into question the availability of information for citizens to make decisions about social, economic, and political affairs. The First Amendment of the Bill of Rights of the United States Constitution resonates with a freedom from government intervention, allowing for freedom of expression and freedom of the press. While publishing has now morphed from Colonial America's penny presses, to niche blogs, multimedia displays, and converged mobile media, the compelling need to oppose government intervention remains as vital today as when the citizens of Colonial America fought for freedom from government regulation exerted by England.

In today's interconnected global world, the ability to share information in a transparent way allows individuals to hold the world open for conversation. In an unencumbered information marketplace, citizens of the world can partake of views that harmonize with their existing viewpoints, learn about opposing points of view, and engage in a robust exchange to debate concepts to arrive at consensus of understanding on a societal level. Further, the individual operating in this transparent realm experiences an "openness" that may curtail traditional notions of privacy, giving the individual a newfound access to an authentic projection of one's self. With the failing trust on a global basis, authentic sharing of information provides a viable recourse. Yet government intervention and private manipulation of data warp the marketplace. Transparency sheds light into corruption and can prevent obfuscation from the facts. Communication plays a vital role in this process. Further, the

Digital Age demands that we consider information dissemination methods including online search engines.

The right to be forgotten, while it is celebrated in Europe as a crusade for privacy, propels a concealment that can obscure truths and warp perceptions. Amid nascent efforts to remove data from the World Wide Web, promote online security, and create Internet governance protocols, society drifts away from a culture of "forgetting" to one of perpetual memory, giving an archival nature to personal information. The effective concealment of discreet personal facts becomes impossible in the digital era, breaking the information dam to give way to a watershed of information. In a work on memory, scholar Paul Ricouer (2004) has written, "To put it bluntly, we have nothing better than memory to signify that something has taken place, has occurred, has happened before we declare that we remember it" (p. 21). Ricouer states that memory operates with "multiplicity" and varying degrees of distinctness. In the Digital Age, memory is now aided by technology and the permanence of information that can be found on the World Wide Web. This permanence signifies a departure of a society that once operated with a frail memory that willingly embraced forgiveness by virtue of simple "forgetting." Even Europeans who have petitioned Google with takedown requests may find that the information lingers on servers and industrious researchers can discern the names of individuals who have asked for the links to be removed (Scott, 2016).

So the conundrum remains: How do governments balance freedom of information with digital governance and movements for the right to be forgotten? Alternative approaches have been suggested, ranging from the creation of new public laws, the publication of catalogs with certain outcomes, and the institution of case-by-case decision making (Lee, 2016b). Yet in the United States, the creation of new laws to encumber freedom of expression seem implausible. Instead, United States jurisprudence defaults to existing laws of speech regulation in the context of obscenity, libel, national security, and freedom of information. While European Union countries attempt to make sense of the right to be forgotten and incorporate more stringent aspects of data protection, the United States stands in a position of observation and contemplation about the implications of data removal on a worldwide basis. In the current environment characterized by the ubiquity of online public records, social networks and global information sharing, Nissenbaum presents the notion of "contextual integrity," a situation that considers the roles, activities and norms to contemplate a new understanding of privacy in the Digital Age (Nissenbaum, 2010). Calling the existing conceptions of privacy "unable to adapt to the new landscape," Nissenbaum prescribes a new methodology for privacy that examines information flow within context and may offer insights for Google and other search engines grappling with the full implementation of the right to be forgotten in Europe. However, societal fear and preoccupation with privacy should not eclipse concerns about the intrusion of government into information transmission and the sharing of ideas among citizens.

A new generation of information producers, content generators, and social media key influencers proliferate the total inventory of words and images found online today. Yet these content producers remain devoid of any code of ethics or adherence to values of truth and accuracy. While the marketplace becomes cluttered and complicated for the viewer, the marketplace must stand and remain protected under these circumstances so that unrepresented voices can gain access and the citizenry can form consensus. By allowing for authentic transparency and avoiding an Internet with holes, contemporary communicators can strive to keep the marketplace open, pushing prevailing ideas to the forefront. The Seven Tenets for the Digital Age send a rallying cry for all individuals with influence, authority, and the communication savvy to preserve the integrity of the First Amendment, amid the changing dynamics of information sharing and data preservation or retrieval in the Digital Age. Encumbrances on the adequate flow of information, with data removed and memory holes ripped into the fabric of society's recollections of past events, lead to a far more deleterious picture.

With the specter of government intervention, we have no other choice but to hold the search engines open for information distribution. The question of one government's ability to erase history and prevent the open publication of information in global democratic societies darkens the ability for global collaborations. We may ask ourselves, "What gives the European Union the power to decide what is fit to appear in global search engines by requiring the compliance of private enterprise entities like Google, Bing, and others to de-link information that has already been printed?" In light of these changing circumstances, perhaps United States citizens may unite behind the Free Speech Manifesto for the Digital Age and frame thinking to consider the corresponding impact of data removal, censorship, and the warping of the treatment of historic events.

THE FREE SPEECH MANIFESTO FOR THE DIGITAL AGE: SEVEN TENETS TO PRESERVE INFORMATION FLOW IN DEMOCRACY

1. Uphold the First Amendment to preserve democracy

The European right to be forgotten is a violation of the First Amendment of the Bill of Rights of the United States Constitution that guarantees freedom

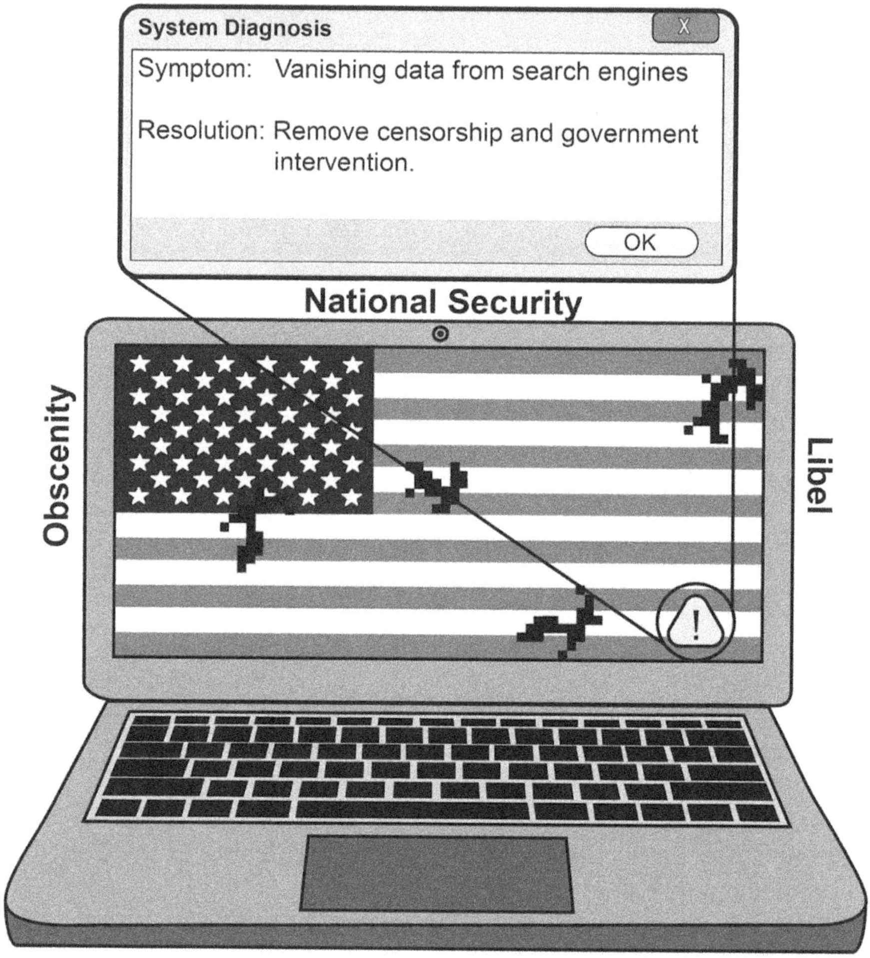

Figure 7.1: Vanishing Data
***Source*: Figure used with permission of Marisa Peters.**

of speech and freedom of the press. European efforts to eradicate data from search engine results through de-linking violate the spirit of the constitutional framers of the United States. American citizens must question a Europe-centered digital governance move that impairs freedom of information. The unbridled access to information undergoes threat by the European General Data Protection Regulation.

2. Safeguard journalism with data availability

The right to be forgotten creates memory holes in the World Wide Web. When data is de-linked from a common search engine, the aspect of discoverability is diminished, debilitating searches conducted by journalists and private citizens. Transparency of government and access to information serve as hallmarks of information flow in democracy. The ability for constituencies to locate information, observe societal happenings, and proffer further information remains paramount in the Digital Age. The erasure of information from online search engines through de-linking casts a pallor over the free flow of information globally. Journalists will face a handicap that will only be overcome through productive, innovative, and investigative means. Blockades to accurate information may prompt subversive behaviors, including hacking, information-leaking, and other means in efforts to preserve transparency for citizens who seek the opportunity for informed decision-making.

3. Preserve historical memory

The right to be forgotten changes history and warps perceptions of past events and societal interpretations. This legal mandate in effect punctures data flow in the recollection of historical events related to a person, social occurrence, or political movement. By obscuring the historical reflections that can occur, the ability for individuals to vet political candidates becomes flawed. This maturation of the Information Age presents a new circumstance in which information-rich nations denote memory. In the information-rich society, the Digital Age demands a societal shift toward an "all remembering" society away from a society of forgetting.

4. Reject the conflation of privacy arguments with First Amendment protections

As United States citizens, we must reject the argument that information removal from the web is a privacy concern. European citizens who seek the de-linking of information assert a privacy argument that claims an individual should have the right to remove data that is outdated or no longer relevant. Yet, de-linking diminishes the discoverability of information, leading to concealment and secrecy. The information presented online becomes published as part of the digital archive. Libel actions still provide legal recourses for citizens seeking relief. A tension between privacy activists and advocates of freedom of information will continue until the notions of privacy and de-linking are unbundled.

5. Preserve the marketplace of ideas in the digital archive

Democracy cannot thrive without an unencumbered and robust exchange of ideas. The right to be forgotten corrupts the marketplace of ideas, as data may be removed through de-linking, along with accurate portrayals of past historical events, social occurrences, and political movements. Individuals cannot see what is no longer there. Transparency becomes corrupt. Nascent regulators operating in the name of individual privacy make a grave mistake, sacrificing the free flow of information in society in exchange for government intervention into information availability.

6. Resist power shifts to private organizations with censorship power over information availability

The inculcation of power to Google by the European Union bestows a vast power to the for-profit enterprise—to determine the availability of information commonly found on the Internet. The company dropped the "Don't be Evil" motto in 2015 and today it remains at the center of litigation tugging between First Amendment protections and privacy provisions in Europe. Google's role as the web's "decider" makes them a proxy for government control over information. While journalists and media outlets have long provided a "gatekeeping" role for information in democracy, today's Internet search engines were not created to function as an agent of editorial control over information. Rather, the search engines have served as a conduit for information and as a for-profit enterprise that accepts advertising revenues to escalate search engine results. The effective empowerment of private search engines by the European Union takes effect in the context of the European General Data Protection Regulation, yet American citizens and information scholars must resist the shift to give censorship power to these organizations.

7. Guard against global hegemony as the European Union exerts power over the availability of information found on the World Wide Web

United States citizens who learn about the right to be forgotten may become baffled to find out that a private censor is de-linking information across multiple domains in Europe. As American citizens, we acknowledge the ease of access to information and recognize that nondemocratic governments exert draconian measures on information flow to quell rebellion and innovative thinking. While the First Amendment seemingly protects the U.S. Google domain as of now, the international composition of information changes in the right to be forgotten environment. United States citizens and First

Amendment scholars must guard against these hegemonic moves that may artificially adapt thinking and perceptions of the world around us.

THE FUTURE FREE FLOW OF INFORMATION

The digital picture of the future will look more like a colorful mosaic that one created with a fluid brushstroke. Small tiles of information will be plucked from the portrait, at times obscuring the picture and warping the collective memory. In May 2018, all member states of the European Union must comply with the European General Data Protection Regulation. This date marks a new era of information management along all search engines as private data collectors like Google become "deciders" of what is viewed on the World Wide Web. This does not bode well for the free flow of information in society. Democracies in particular may suffer as citizens no longer actively seek out information, but rather recline in passivity, allowing individuals to shape their own online persona through the select de-linking of "outdated" or "no longer relevant" information, making the vetting of individuals for public office a daunting charade.

Through a thoughtful intellectual response to forces coming across the pond from Europe, scholars in the United States must simultaneously consider digital governance within the broader context of the free flow of information. The elastic nature of the First Amendment has endured through generations of disruptive technology innovations. The fundamental freedoms and protection from government intervention found in the Bill of Rights remain strong enough to combat search engine manipulation. The free flow of information in the Digital Age may be protected and serve a meaningful purpose in the world's democracies. Yet, the inexorable link between democracy and the free flow of information is a delicate force that must be nurtured by a free media, freedom from censorship, and absence of government intervention. Encumbrances on journalists, free speech activists, citizen speakers, and the eradication of information because it is "no longer relevant" present a threat to the marketplace of ideas and obscure information transparency.

In the United States we must counter movements toward global information hegemony, because we cannot allow a trampling on freedom of speech in the United States today. We must take the first step to aptly recognize search engines are private companies given vast power by the European Union. Google and other worldwide search engines possess virtual ability to determine the availability of information both in the United States and Europe, transcending legal boundaries and the jurisprudence of Europe and the United States. Yet today, we operate among a failed consensus on how to handle this information. Further, we must put the privacy argument aside, and perhaps put it to rest permanently. The gravity of the preservation of collective memory and the need for transparency in economic, social, and political models remains a stalwart need in the Digital Age. Technology in this era elongates society's memory, yet the availability of information must be protected so we can reflect on the past and make decisions for the present and future.

Chapter 8

The Preferred Position of Freedom of Information

Amid the cacophony of concerns over privacy, data retention, and personal identity, the overriding need to protect the First Amendment of the United States remains imperative, because freedom of speech and historical accuracy perform integral roles in the practice of contemporary democracy. We may need to look no further than the preferred position doctrine of the First Amendment to frame the discussion of the right to be forgotten in the United States. While this doctrine itself has waxed and waned throughout history, the preference of constitutional freedoms above others remains an understandable principle for the Digital Age. Therefore, the fundamental constitutional right of freedom of expression trumps other rights claimed under the right to be forgotten and similar movements that diminish the availability of free access to information.

ACCEPTING THE PREFERRED POSITION DOCTRINE OF THE FIRST AMENDMENT IN THE U.S. CONSTITUTION

While some citizens may balk at placing the constitutional rights of freedom of speech and press contained in the First Amendment above other rights, the preferred position doctrine suggests the necessity of doing so, and, in the postmodern Information Age, this right may have escalated to a new level of prominence. The entanglement of freedom of speech in a democracy, combined with the distribution mechanism of the online sphere, suggests we must look back to the preferred position doctrine to look forward to future healthy democracies.

The preferred position doctrine prescribes a judicial standard that presents a hierarchy of rights in which constitutional freedoms gain more "protection" than other rights. The preferred positon doctrine, which is also referred to as the "preferred freedoms doctrine" is often traced to U.S. Supreme Court Justice Oliver Wendell Holmes, who recognized the primacy of the First Amendment in *Schenck v. United States* (1919). The concept was later advanced by Justice Benjamin Cardozo in *Palko v. Connecticut* (1937), suggesting that certain fundamental rights like freedom of speech, religion, and the press are the very essence of a scheme of ordered liberty (Palko 1937). Later, the notion of ordering rights in a "preferred position" arrived in a footnote of a majority decision in the *United States v. Carolene Products* case (1938), whereby U.S. Justice Harlan Fiske Stone argued that civil rights and individual liberties merit a "preferred position." However, in the history of American jurisprudence, invoking the preferred position and ordering of rights gained favor in the Warren Court of the 1950s and 1960s, and the Burger Court of the 1980s (Pacelle, 2017).

CONTEMPLATING THE ROLE OF GOVERNMENT, CITIZENS, AND PRIVATE ENTERPRISE

The First Amendment's freedom of speech clause has retained an elevated position in contemplation of the preferred position. The threats from the European right to be forgotten now indicate that preferencing of this right may fall, not as the result of a change in the American jurisprudence, but rather as a deleterious side effect of the European General Data Protection Regulation. While U.S. citizens venerate the right, European audiences do not understand or interpret the protection in the same way. With global web governance, governments and private companies must strike a balance between these conflicting viewpoints.

For example, Europeans tolerate enhanced levels of government intervention, while in the United States, citizens become squeamish about elevating government's role and allowing for undue intrusion in our affairs. From a mass communication perspective, though, the United States regulates television and radio, and invokes legal regulation in the cases of libel, threats to national security, and obscenity. The aftermath of 9/11 in the United States, resulting in the collapse of New York City's Twin Towers, has ushered in an era that embraces new thresholds of tolerance for surveillance. The USA Patriot Act: Preserving Life and Liberty (2001) allowed for enhanced levels of surveillance and intergovernmental agency collaboration that could lead to more rapid detection of terrorist activities. The incorporation of many global technology companies, including Google and Facebook in the United States, suggests the federal government may have the opportunity to enforce regulations that guide digital freedom of information.

The Role of Government to Balance Freedom of Information

Government officials in the United States must contemplate an appropriate balance between government regulation and freedom of expression. Europe has taken a broad-sweeping approach to data protection and regulation that brings a new standard to global information retention. With the May 25, 2018, enforcement of the European General Data Protection Regulation, multinational corporations scramble to insert proper policies, required technologies, and operational procedures to comply with the regulation or face fines and other negative consequences. The government's role has expanded to regulate private enterprise in this arena, in a fashion similar to the licensing and regulation of U.S media such as television and radio broadcasting. In the global information society and in our respective countries, we must contemplate the role of government as a vehicle to protect against information monopolies and censorship, while the government simultaneously protects the environment for the proliferation of information through proper web governance, data retention, and information retrieval. This is a tall order. The European Union has set forth a best effort in the General Data Protection Act and sought to satisfy multiple dimensions of privacy, notably the right to be forgotten. The classification of Google as a data controller under the law, however, puts that company into a powerful position that encompasses both responsibility and control over the availability of information found through the well-embraced search engine.

The Role of the Citizen and the Media Consumption Test

The citizen's role as an audience member becomes more important in the Digital Age as media sources fragment. Citizens become marketplaces of one, seeking information that interests them. The Digital Age presents new circumstances that place the freedom of speech and press clauses of the First Amendment under siege. Today, citizens in both the United States and Europe observe an environment where the term "fake news" has become a household word. However, does the average person understand the definition of fake news, and if it is, in fact, really fake? The trouble with assuming fake news is "false" presumes that the facts have been checked and that a third party has discredited the fake news, based on some other objective source of facts. Instead, myths and misperceptions about the believability of news stories becomes convoluted in the Digital Age, reducing the citizenry to a state of confusion and bewilderment about what is fact, what is fiction, and what can be discarded as hearsay or falsity. The confounding predicament worsens when an authoritative figure disseminates the false messages. Even for the discerning eye, this can become challenging. Today, thinking individuals must consider the source of information and psychologically run though a mental checklist for validity, authenticity and truth. This mental checklist may be considered as a media consumption test.

Media Consumption Test (Five-Prong Assessment)

1. Is this source believable?

The reader first seeks to discern if this is a believable source. Does the source present independently verified infor mation? Is it a true news source or an opinion source with a particular agenda?

2. Does this news story harmonize with my beliefs?

If this news story makes you shake your head in affirmation, perhaps you operate in the so-called "echo chamber" and find yourself operating in a system of homogeneity of thought and action strikingly similar to your peer set and family.

3. Will I seek out other viewpoints that may contradict my own viewpoint?

To effectively break away from the "echo chamber" and learn about other perspectives on political, social, and economic issues, the individual may purposefully seek out other perspectives that enlighten the viewpoint.

4. Am I willing to accept this news story as "truth" or do I classify this as "fake news"?

Immediately after exposure to media content, the citizen exercises a decision over personal belief patterns. The person may choose to accept the story as "truth" and memorialize the information and potentially into a personal belief system, or they dismiss the information as fake news and thus discard the information for failing to pass the assessment of accuracy and truth.

5. Will I share this news with my friends on my social media networks?

In the ubiquitous atmosphere of digital content co-creation, individuals undertake an activist role and become champions of specific political viewpoints, causes, or commercial enterprises. The individual takes a final action step by deciding whether or not to share the information with another person.

This mental checklist for media consumption presents a lot of hard work for individuals bombarded with multiple messages from various technologies and distribution mechanisms. For example, the reader may designate a CNN news alert that streams across the mobile phone more reliable than the social media post relayed from a family friend with a partisan ax to grind. For some, separating tenable information from the unbelievable seems a worthy exercise. For others, though, the higher level of investigation into mass media messages never occurs, and the messages may escalate to "facts" without the benefit of the independent verification of truth or the objective collaboration of multiple news sources. Instead, many citizens fail to question the source and believability, taking a short cut directly to, "It must be true, I read it on the Internet." Thus, we may wonder, what does this activity say about the health of the marketplace of ideas? Mass media scholars must carefully examine the marketplace to check the health of the information exchange amid the cacophony of information presented. Some authors present information to advance in a benevolent way, while others use more forceful aspects of persuasion. Individuals who share information seek to influence social, economic, and political circumstances and enhance their own credibility as an opinion leader. Information sharing occurs in an environment of regulation. The audience possesses a new-found responsibility to discern truth by discarding falsities and untenable beliefs. Just as a juror listens to multiple testimonies, the contemporary media consumer engages in the broader world through multimedia interactions that inject news and information into their daily lives, forcing each person to accept or reject notions to believe or not to believe.

The Role of Private Corporations and Private Industry

The European General Data Protection Regulation gives Google and others search engines an elevated, influential level for curating information. Further, the regulation provides stipulations about data retention and the ability for citizens to petition other corporations to remove information from data warehouses, forcing private industry into a new regulatory environment that can lead to fines and other penalties for compliance failures. By looking inward to private corporations and private industry, some management teams may heed the call of ethics and professionalism to allow for prudent management of information and responsible stewardship of data in the Digital Age. In 2017, Facebook attorneys underwent questioning by a United States Senate panel about the acceptance of paid advertisements during the 2016 Presidential election in the United States (Pullen 2017). This situation evokes many questions: (1) Does World Wide Web advertising for elections for political office require the same governance as disclosures found in print and broadcast advertising paid for by special-interest groups? (2) Does the ability of a foreign government to place persuasive advertising in a prominent national social media channel present an ethical dilemma for the social media company? (3) Is there a legal obligation for the social media channel to report the solicitation for the advertisements to federal authorities, including the Federal Election Commission? This situation and similar predicaments that media enterprise find themselves in represent the new landscape for technology companies existing in modern democracies, as politicians and their advocates attempt to use all available means to persuade voters and promote a predetermined, desired outcome. A discriminating viewer may distinguish between journalism and paid media content in the form of advertisements. Facebook and other social media outlets are for-profit enterprises. However, the failure of the company to report the use of a foreign government illegally making payments for politically advertisements to the Federal Election Commission remains questionable. Movements that press for government regulation in World Wide Web governance, data retention, and control over the world's information must undergo a continual counterbalance, with the value placed on freedom of information in society. This is an ongoing dialogue that encompasses political speech, ethical boundaries, and circumspection about new technologies that allow for communication to the masses.

THE FIRST AMENDMENT UNDER SIEGE

The prevalence of the term "fake news" reflects the postmodern Information Age that encompasses changing dynamics for media organizations, content creators, journalists, and citizens attempting to function in society and in their respective democracies. Unfortunately for the typical media consumer, confusion erupts as the ability to classify information as truth or mistruth confounds the average consumer. In this new era of bewilderment, pro-active communicators may easily dupe the consumer, prompting the reader to give up altogether on the believability of news and information. The decline in the believability occurs amid an atmosphere of eroding objectivity in the practice of news reporting and of the emergence of partisan media that vocalize specific agendas. In this environment, citizens may seek out information that harmonizes only with their existing belief system and they may fail to gain access to countering viewpoints. This sets the state for homogeneity of thought, whereby a propagandist may set the agenda and the citizen advances it by sharing the information. In tandem with decline of objective journalism, we observe the deterioration of the Gatekeeping Theory of communication. While editors and producers have traditionally performed a decision-making role that determines the appropriateness and relevancy of information published, based on the newsworthiness and the independent verification of the facts, citizen publishers today thrust the gate wide open to advance opinion and mistruths masqueraded as facts.

We simultaneously observe the vilification of certain media figures and the regaling of others, boosting them to popular celebrities that wield phenomenal power to persuade vast numbers of followers. The postmodern Information Society does not present a level playing field and articulate social media leaders gain real influence through the power of persuasion. The ability for political candidates or social causes to gain traction does not necessarily occur door to door any more, but rather travels through technology in nanoseconds, giving citizens that ability to tap into concerns geographically both near and far. The new global communication channels, fueled by technology and citizen journalism, present new challenges for the First Amendment and suggest that citizens remain baffled by the scope of First Amendment freedoms and by the limitations on communication for libel and national security. Some vocal advocates may invoke the freedom as a blanket right to say whatever they want, whenever they want, to whomever they want. Despite this arbitrary application of the First Amendment, freedom of speech and freedom of the press remain constitutional rights guaranteed by the Bill of Rights and therefore must gain a preferred position, as the technology landscape foretells changing media circumstances for the flow of information in the Digital Age.

Chapter 9

Defining Freedom of Information in the Digital Age

While the European right to be forgotten guides practices in Europe, the movement by the European Union prompts American jurisprudence to truly consider the application of the First Amendment to the web, unearthing issues related to World Wide Web governance and the regulation of search engines as "data controllers." We must now ask ourselves if search engine results are, in fact, protected speech, or are they an algorithmically based work product like a car alarm or motion-sensor doorbell? Scholars today do not dispute that as "data controllers" operating in the digital environment, Google and other search engines have ascended to an ultra-powerful position, but the legal implications of this global regulation still remain opaque. Pressing dilemmas about the availability of online information and the specter of European hegemony means we must take pause and consider the consequences. Underscoring the idea that valuable institutions such as democracy, liberty, and pluralism depend on the fact our lives cannot be completely controlled, we are warned against the acceptance of an information "kill switch" (Gstrein, 2015).

Data controllers in Europe now subject to the right to be forgotten face the daunting challenge of reviewing thousands of requests to remove information allegedly no longer deemed relevant. Scholars suggest similar implementation in the United States seems unlikely. While controversy over the right includes a discussion of the multicultural understandings of privacy, the U.S. constitutional right of freedom of expression in the First Amendment tips the balance in favor of the free flow of information in society. U.S. citizens, beholden to values of liberty and protection from government intervention, vehemently argue the right to be forgotten precludes freedom of access to information and can lead to information censorship in society. Vigilance of new technologies and respect for culture differences are required to develop workable solutions for the right to be forgotten (Bennett, 2012; Ambrose & Ausloos, 2013). By casting privacy arguments aside, including ones that seem to suggest an unconstitutional statutory construction of regulation and those focusing on U.S. constitutional theories of the First Amendment, the argument strengthens the opposition against the right to be forgotten. Do citizens living in a democracy welcome World Wide Web censorship and the infringement of a foreign government in determining what appears on search engine results? This movement suggests a chipping away at a fundamental First Amendment right embodied in freedom of speech that could erode further if a right to be forgotten emerged in the United States. American citizens have little recourse for the encumbrances on the information in Europe, except to once again exercise free speech in protest of the right to be forgotten, opting instead for an open, transparent information atmosphere.

This book began with an exploration of the First Amendment, by examining the history of the Bill of Rights and the exacting verbiage carefully crafted to protect freedom of speech, freedom of the press, and the right of the people peaceably to assemble. Arguments presented in this text assert that the European right to be forgotten damages these freedoms. Regulation can imperil freedom of speech through arbitrary censorship. Freedom of the press undergoes Digital Age duress as memory holes limit the ability to find information on political candidates and social happenings. Lastly, the right of the people to assemble may fail to come together, if information disappears under a right to be forgotten request. The adherence to a free media environment that encourages public debate remains core to U.S. democratic principles. The elasticity of the First Amendment continues to broaden as new modes of communication technology emerge, and the legal protection still embraces the freedom of information principles woven in the cultural fabric of the United States. Chapter 1 provided a historic overview of the First Amendment within the context of the current right to be forgotten landscape and set the stage for chapter 2's discussion of the marketplace of ideas, a bedrock theoretical principle that finds its way through U.S. economic and political principles. To preserve the integrity of the marketplace, information must remain available and accessible by the citizenry. Today the ubiquity of the Google search engine has created a de facto monopoly over information control. Google already created a subjective ranking of search engine results (utilizing page rank) and the company participates in the full array of commercialism, accepting advertising fees for preferential treatment on the search engine results. In conjunction with their reliance on the online search engines for information access, Digital Age citizens often approach the search engine with a belief that the search will yield valid results. While this may be considered naïve, we seem to operate as if we have a covenant with the search engine, trusting that we enter search engine terms and the search engine serves up the complete results. In the opaqueness of the right to be forgotten, that covenant with the search engine changes. Trusting the search engine results as an accurate reflection of the total body of information about a chosen topic or individual no longer holds true. Instead, holes created by missing information obscure the total picture and can warp our perceptions of individuals and political candidates in the contemporary world. In chapter 3, readers learned about the connection between freedom of information in a democracy and the traditional watchdog role of journalism that empowers reporters to keep an eye of scrutiny on public institutions and citizens who enter the publish sphere to run for public office in a democracy. Specific journalistic protections, including the Freedom of Information Act that gives reporters access to federal documents and the Government in Sunshine Act that ensures journalistic access for public meetings, stand as evidence of the cultural and legal fabric that supports the First Amendment in the United States. The presence of these journalistic protections further underscores the need for transparency in the Digital Age and reveals that the nexus between freedom of information and elected officials remains a zone meriting protection to ensure accurate reporting and information-sharing with citizens.

The impact of the European legal notion of the right to be forgotten on online search engines gives rise to a myriad of legal and constitutional issues for the United States. The enforcement of the European General Data Protection Regulation manifests serious questions about the macro issues of web data, privacy, and the role of communication professionals in proliferating atmospheres of free flow of information. The compelling need to maintain digital privacy through the control over personal information dominates privacy discussions. This conflict creates a conundrum for journalists and professional communicators because the movement could lead to information censorship and threaten core values. The Google search engine crosses international borders, complicating the administration of the right to be forgotten in the United States. For journalists and public relations professionals, it is relevant to consider the active free flow of information within the context of dialogic ethics. The commonly held two way-symmetrical theory of public relations (Grunig & Grunig, 1992) aligns with dialogic ethics, allowing public relations professionals to build relationships with key publics. This dialogue may be interrupted by the right to be forgotten. This book asserts efforts to censor and de-link search engine results inhibits the free flow of information in society. First Amendment scholars critique the right to be forgotten as unconstitutional by virtue of the Bill of Rights, citing First Amendment constitutional foundations such as the Marketplace of Ideas Theory, the Meiklejohnian Theory, and the Absolutist Theory as proof points for the repudiation. Professionals in public relations and journalism question aspects of government intervention and arbitrary information control. In what have been coined as "deciders" (Rosen, 2012), the workers at Google inherit great power over information management. The consolidation of control with the search engine leaves the United States adrift from the fundamental principles of the free flow of information integral to the practice of democracy. Further, in a right to be forgotten environment, the global World Wide Web loses freedom and openness. Moving forward, the transnational legal and economic issues must be addressed in the context of global standards, which may not be compatible when attempting to balance the value of privacy rights in Europe with the protection of free speech rights in the United States.

The concept of privacy rights remains central to the discussion of the right to be forgotten. European citizens appear at times to attach greater value to digital privacy than residents of the United States, who often place freedom of expression in a preferred position to privacy. The European General Data Protection Regulation prescribes changes for its member countries, and, in doing so, has activated similar calls for data protection in the United States. However, parallel actions in the U.S. remain stifled by an adherence to the First Amendment. Pitfalls populate the path forward for the United States. Vigilance of new technologies and respect for cultural differences are required to develop workable solutions for the right to be forgotten (Bennett, 2012; Ambrose & Ausloos, 2013). While information passes through natural life cycles and matures through the simple passage of time, the right to be forgotten intervenes in the natural digital decay resulting from a multitude of reasons such as hardware failure or software errors. Societal issues of Big Data retention and information preservation will continue to dominate discussions of freedom of speech in the United States and abroad. While it has been stated that the World Wide Web is not a library (Ambrose & Ausloos, 2013), it remains a valuable tool for citizens engaged in a complex contemporary society struggling to understand economic, social, and political circumstances. For information creators, including journalists and public relations professionals, the marketplace of ideas functions in a democracy and must stay open to allow for information disclosure and robust debate. The proactive approach of "de-linking" information from the search engine makes information more difficult to find, placing a greater onus on journalists to find information in an old-economy way, digging through papers in government archives.

The specter of government intervention into the free flow of information stands as a formidable force in the Digital Age, not only in the United States, but also with governments abroad, and concepts of web governance and the ability to regulate content arise. We can look back to the historical roots of the First Amendment and the call for protection from government intervention. Academic discussions erupt about whether or not search engines results are, in fact, protected by the First Amendment. Scholars and practitioners will continue to debate these factors as they delve deeper into the role of algorithms, the exertion of editorial control, and the European classification of Google as a data controller. This text explores an ardent rejection of the European right to be forgotten in the United States. While scholars intellectually evaluate workable parallels in the United States, they look to aspects of forgiveness already found in United States jurisprudence, like the Fair Credit Reporting Act. Yet, these movements to sculpt a modicum of forgiveness through a right to be forgotten parallel in the United States skip over the constitutional imperatives to protect freedom of speech and freedom of the press in our country. We may take notes from historical treatment of the free flow of information in totalitarian governments in an effort to discern the "slippery slope" of government-mandated information control. We can look to Hannah Arendt's "lying world order," and we can ask ourselves if the right to be forgotten robs us of the ability to bear witness to the social, economic, and political circumstances of our time. The global tides of information regulation and data governance across geopolitical lines will daunt contemporary and future political leaders. The confluence of various constitutional models, cultural traditions, and levels of technology mastery provide no simple roadmap for managing information. In the United States, we have traditionally resisted government intervention in freedom of information, suggesting an immediate conflict with the European right to be forgotten. Thus, we must consider the notions of memory, both on an individual and collective society, and evaluate the meaning of privacy in the world of mobile media and the cluttered social media environment.

We find ourselves in an atmosphere that requires us to contemplate memory concurrently on individual and societal levels. The permanence of information shared on the World Wide Web gives human beings an extended "archive" of events that happen from the moment a social media account debuts to the present day. Despite the fact that all information is not favorable or consistent with our "personal brand" or grown-up character, eradication of the information by a search engine counters the legal protection of the First Amendment and contradicts the heritage of freedom of information in the United States. On a societal level, movements to distort history through the a la carte removal of information stand as a worrisome dimension of the European right to be forgotten. Reminiscent of George Orwell's *1984* that articulated the Ministry of Truth's destruction of records and recreation of reality, we may take note from this literature and determine if the right to be forgotten may be doing the same thing to the global information society as the Orwellian memory holes. The ability for actors to manipulate history though omission of information can lead to what Hannah Arendt calls a "lying world order" that fortifies conditions for the rise of totalitarian regimes. Today, web archives remain tied to the management of data warehouses that store the information to be retrieved by search engine users. The conceptualization, function, and governance of the societal archive remain a pressing area of research, perhaps enlightened by the Derridean notion of Archive Fever that considers both the commencement and commandment dimensions of information, whereby both history commences and social order gains organization. We live in an era of a societal turn to constant remembering versus the traditional culture of "forgetting" as a new sense of digital "permanence" replaces the forgetfulness of old economy record-keeping and the natural decline of human memory moving through time.

With the constant gaze of technology, the construct of human privacy changes. The projection of human actions onto a digital archive opens citizens to a new realm where information is constantly available despite the passage of time or the evolution of one's own maturity. While the European right to be forgotten attempts to take a stand to help individuals "forget their own past" by righting past wrongs through removal of information, we cannot accept a prima facie privacy argument. Instead, we can understand how it happened, considering the cultural attachment to privacy found in Europe and the inherent distrust of government found in the United States. However, this text asserts that privacy activists misappropriate the privacy argument, and the focus must remain on the protection of freedom of information. Societally, we may therefore take pause and incorporate human ethics and consider the "who" in the equation for monitoring privacy. Does liability for privacy protection reside with the government or individuals advocating on their own behalf? These complicated questions will continue to conjure up conflict, while technology innovations contribute to more societal surveillance, and Big Data storage, and predictive analytics usher in new techniques for data use that have not been conceived as of this writing. In a reconciliation of the privacy rights and freedom of information, we can rapidly conclude the removal of information that has been previously published and is true presents an awkward and scary proposition for modern democracies. Every day Google receives requests under the right to be forgotten and must process accordingly or face legal sanctions. While the United States does not have a similar right to be forgotten provision, pressures to formulate some type of comparable recourse for removal of "outdated" and "no longer relevant" information remain. This creates a vast rift between individuals who want to protect the free flow of information in society and uphold the First Amendment, and others who are willing to sacrifice a constitutional protection that is foundational to democracy in exchange for a short-term "take down" of information. The removal does not just signal the end to an uncomfortable past for someone; it creates an environment whereby memory can be distorted, history re-created, and society becomes cast adrift wondering what is true and what is false, attempting to navigate through an Arendtian "lying world order." Considering the horizon of web governance and global information security models, the opportunity to adhere to First Amendment principles remains.

This text suggests a "First Amendment Manifesto for the Digital Age" holding the world open to information exchange and giving citizens in the United States access to a marketplace of ideas that is not artificially tampered with through the erasure of information. The directive suggests principles including: (1) Uphold the First Amendment to preserve democracy. (2) Safeguard journalism with data availability. (3) Preserve historical memory. (4) Reject the conflation of privacy arguments with First Amendment protections. (5) Preserve the marketplace of ideas in the digital world. (6) Resist power shifts to private organizations with power over information availability. (7) Guard against global hegemony as the European Union exerts power over the availability of information found on the Internet. This manifesto creates a rallying point for individuals who honor the First Amendment and its role in the health of democracy, and a recognition that infringement on ideas may contribute to deleterious consequences for the electorate and the democracy itself.

The preferred position doctrine is examined in chapter 8, enlightening discussion of the status of this constitutional freedom in contrast to other legal rights. The way we think about the First Amendment may find a new light because of the changing circumstances. Albeit, criticism of the marketplace of ideas theory occurs when forces for idea transactions and marketplace regulation come to bear (Blocher, 2008). Further, while the marketplace of ideas functions as a theoretical metaphor when contemplating First Amendment protection, unfair competition of ideas and imperfect information can obstruct the performance of the marketplace (Brazeal, 2011). So how do Digital Age citizens in representative democracies proceed? The European legal notion of the right to be forgotten on online search engines unearths a myriad of legal and constitutional issues for the United States and conjures serious questions about the macro issues of Internet data, privacy, and the role of the information seekers. A tug of war blooms throughout the globe between privacy activists, who seek the ability to have information de-linked, and First Amendment advocates, who claim search engine removal policies "rewrite history" and lead to arbitrary censorship. Scholars continue to grapple with defining a common terminology for the legal right and the use of terms such as oblivion, deletion, and de-listing (Forde, 2015). We live in a formative era for understanding the right to be forgotten and its impact on global information availability. The Google search engine crosses international borders, complicating the administration of the right to be forgotten in the United States while privacy activists urge federal authorities to apply similar standards in the United States. By examining constitutional arguments, including First Amendment theories such as the Marketplace of Ideas Theory, the Meiklejohnian Theory, and the Absolutist Theory, we can begin to see that the right to be forgotten encounters formidable opposition in the United States.

For U.S. citizens, legal analysts, Digital Age practitioners, and scholars of media and communication, it is now time to thoughtfully consider the marketplace ideas as a vital foundation for democracy and "hold the world open" for information disclosure and healthy debate. Democratic countries that value freedom of expression may reframe discussions of privacy and protect the fundamental tenets of the free flow of information in further examination of the right to be forgotten. With the advent of Big Data and interference with an individual's privacy, the notion of the right to be forgotten should gain consideration in a wider context that contemplates mass surveillance and data processing (de Rosnay & Guadamuz, 2016). The right to be forgotten is not compatible with the constitutional protection of the First Amendment. Despite efforts by privacy activists to enable citizens to take action, the specter of government control lurks behind the right to be forgotten.

IMPACT ON NATIONAL SOVEREIGNTY: EFFECTING CENSORSHIP ACROSS GEOGRAPHIC BORDERS

The ability for the European Union to regulate online data policies that influence the availability of information on the World Wide Web indicates a potential infringement on national sovereignty not only among European countries, but also in the United States which may experience aspects of censorship that fly in the face of the First Amendment. Moreover, the ability for a single individual working at a search engine to make these critical decisions seems arbitrary at best. "While privacy in the new age of information is undoubtedly important, the right to be forgotten harms more than it helps. It allows a single individual to determine what the rest of the world sees, threatening to censor the World Wide Web in its adolescence. It undermines national sovereignty and democratic values by promoting one culture's adherence to privacy over another culture's freedom of expression" (Cunningham, 2016).

Further Global Implications

The right to be forgotten represents a movement of global significance. The procedures for information-sharing and data governance across international borders require significant study, legal collaboration, and an appreciation for various cultural traditions. The incorporation of the right to be forgotten by all members of the European Union requires monitoring and observation.

The planned departure of Great Britain from the European Union leaves the right to be forgotten hanging in a political balance, and scholars call for more thoughtful consideration of information context, the data protection landscape, and freedom of information concerns (Townend, 2017). A study of media coverage of the right to be forgotten yielded insights that a chasm exists between the language used to describe the right to be forgotten in media outlets in the United States and Great Britain, further suggesting that national identity, political orientation, and capitalism play roles in the news coverage treatment (Shahin, 2016). While Google deals with pressures to honor takedown requests across multiple domains, scholars thoughtfully suggest approaches, including Svantesson's (2015) "Model Code for Determining the Geographical Scope of De-Listing under the Right to De-listing" that considers the context for information and a consequence-focused approach.

The European right to be forgotten provides a legal model for other countries attempting to define measures for data removal. Russia enacted a law in 2016 that requires search engine operators to remove information if the information distribution violates the law, is not accurate or factual, or has lost its importance (Andryushchenko, 2016). In the Netherlands, one of the countries of the European Union, the right to be forgotten matured through several judicial decisions before it was mandated through the European General Data Protection Regulation (Verheij, 2016). A Belgian court has ruled that the right to be forgotten is part of the constitutionally guaranteed "right to be left in peace" in one's private life (Ginsberg, 2016). The absence of a consensus of administration across Europe portends a tumultuous time for Google and other search engines legally responsible for implementing the regulation. The application of national sovereignty to control information becomes a challenge in the Digital Age. The presence of subversive and other information outside government auspices continues to grow amid hacking, virtual private networks, and innovative technology tools that utilize satellites and new mobile technologies. The global information society is coming of age with a movement to reduce information in the name of personal privacy under the European General Data Protection Regulation's right to be forgotten.

Is the Marketplace of Ideas Broken?

The contemplation of the right to be forgotten occurs amid the digital revolution and changing dynamics of information flow across borders. Facebook acknowledged the acceptance of advertisements from foreign entities that have meddled with the United States presidential election in 2016 (Shapiro, 2017). This phenomenon suggests the marketplace may be undergoing false market forces, demonstrating a criticism of the economic marketplace and the marketplace of ideas. First Amendment scholars must address ideological thinking about the free flow of information in society and attempt to render a semblance of Digital Age privacy to the individual. U.S. citizens will not simply stand by and let their European counterparts enjoy a higher level of privacy and simultaneously debase the availability of information on search engines. The tolerance for government intervention remains ambiguous, because the United States subscribes to a freedom from government interference, while Europe more willingly permits government management of data issues, as evidenced by the European General Data Protection Regulation.

This book's debut arrives amid the European Union's 2018 enforcement of the General Data Protection Regulation. This denotes a new era of data governance and accountability for data protection, retention, and transmission across Europe. As privacy activists continue to clamor for more government interference in the freedom of information, those in power must consider the large implication of cross-cultural digital governance and the corresponding impact of data removal on the free flow of information. In the United States, a groundswell of constitutional support for the free flow of information must rise. Without the free flow of information, we may not continue to function effectively in a representative democracy. The Constitution's fathers protected citizens against encumbrances to freedom of speech and freedom of the press. For those living and working in the Digital Age in the United States, we must take pause not to allow a foreign government or a private enterprise to compromise the constitutional rights found in the words of the First Amendment. Censorship and the creation of the memory holes cannot endure in healthy democratic societies. Removing the URL from the search engine may not permanently remove the article from the website, but it conceals information. As Digital Age citizens, we demand authentic access to information with a few keystrokes. The information plateau must presume availability, not permeability. In the United States, we must not willingly surrender our wealth of information access by ensuring information does not disappear from search engines the way it is in Europe. Lastly, we may find solace in the ability to work through existing United States legal frameworks for managing data, most notably the body of law devoted to libel. The beating drum for privacy and online reputation management will not likely diminish. Citizens engaged in the larger world want to preserve and protect their persona for maximum personal benefit. We do not need to look to the creation of new laws that chisel away at freedom of speech, but rather we recollect existing laws, including those governing publication of libel and copyright infringement. We live in a time where World Wide Web governance continues to mature, creating a new mosaic of policies and procedures. We now have access to global communication networks that allows for instantaneous communication from one side of the world to the other. What we do with this power remains up to us, as we navigate political economic and social structures to determine how the postmodern Information Age functions for the next generation. The information superhighway runs into the future, yet we must protect the historical accuracy and availability of information to guard against rewriting history and the distortion of reality that could impair the functioning of democracy.

Epilogue

I wrote this book amid an atmosphere of dynamic forces influencing the availability of online information. Today, we notice new breaches of the traditional wall between journalism and paid media. The total availability of information may confound the viewer. The Digital Age information consumer is daunted by constant questions about the believability and authenticity of information. Journalists, once trusted sources because of the objectivity in reporting of information, now work alongside media content producers who capture audiences to advance specific beliefs and falsehoods. As a mass communications professor, I constantly urge students to focus a discerning eye on the information and to ask themselves if the source is true and tenable. While these students pursue academic study of the principles of mass communication, I wonder about the plight of citizens with an untrained eye. These same students encounter roommates, acquaintances, and family members who may not possess the same levels of discernment, thus falling into information grey areas where inaccurate information can be conveyed and take on a life of its own as truth. Myths, rumor, and innuendo spring to the forefront of social media and information transmission, as citizens remain riveted to their smart phones. The failure to "look up" from the smart phone and engage in the larger world through exposure to accurate news and information casts the Digital Age citizen into a vulnerable position. The electorate unable or ill-equipped to properly vet and investigate political candidates contributes to the election of candidates who do not accurately represent the "will of the people." The right to be forgotten represents a sliver of the changing landscape of news and information availability. The compromise of the reverence for freedom of information and protection of media freedoms can contribute to far-reaching social, political, and economic consequences. The toleration for censorship and the creation of memory holes in our collective societal memory will influence short-term and long-term consequences for societal structures.

As a former journalist and practicing public relations professional, I find the European right to be forgotten offends the First Amendment and correspondingly influences freedom of information across the globe. While we live in dynamic circumstances with a loud drum beat for privacy becoming louder in the face of cyber information breaches, the integrity of information must be protected. Data must not only be stored properly, but it must also be available for future generations and accessible for recall so citizens can make conclusions about social, economic, and political policies. The specter of creating memory holes in the World Wide Web and inculcating power into the hands of Google technicians to decide what should be removed from the available inventory suggests that censorship could grow and distort our common understanding of reality. The First Amendment to the U.S. Constitution has endured numerous disruptive forces, yet the contemporary circumstance suggests a new formidable threat to the marketplace of ideas. How the United States handles this issue will paint the portrait of the future and influence the way the next generation exercises freedom of information.

Appendix A

Frequently Asked Questions about the European Right to Be Forgotten

Q: What is the European Right to Be Forgotten?

A: The concept is a legal construct that was recognized by the European Union in 2014, allowing individual Europeans the opportunity to petition search engines for the removal of data that is deemed inaccurate or no longer relevant. Authorized by the European Union, search engines like Google and Bing have honored thousands of requests by European citizens and publish transparency reports about the number of requests.

Q: Why are U.S. citizens concerned about the Right to Be Forgotten?

A: De-linking of information by search engines may violate the U.S tenets of freedom of speech and function as a form of censorship. The right to be forgotten enables the rewriting of history through adaptations of the depiction of contemporary events.

Q: How does the European Right to Be Forgotten violate the U.S. Constitution?

A: Scholars assert the Right to Be Forgotten counters the free speech clause of the First Amendment of the Bill of Rights. The First Amendment states, "Congress shall make no law respecting an establishment of religion, or prohibiting the free exercise thereof; or abridging the freedom of speech, or of the press; or the right of the people peaceably to assemble, and to petition the Government for a redress of grievances." While the European Right to Be Forgotten emanates from a recent legal action taken by the European Union, U.S. citizens with a deep legacy of appreciation for freedom of expression remain wary of government intervention into the free flow of information in society.

Q: Do search engines have the ultimate authority to determine if information is de-linked?

A: Search engines operating in Europe today make a la carte decisions about data removal after an individual has initiated a web form that explains the nature of the request. The search engine, operating as a private enterprise with governmental authorization, maintains the decision-making power about whether to maintain the hyperlink in search engine results or honor the de-linking request.

Q: Is the government mandating the Right to Be Forgotten?

A: The European Right to Be Forgotten is a mandate from the European Union. The United States government has not instituted a right to be forgotten for search engines to de-link information found false or no longer relevant.

Q: Can journalists still gain access to information that has been de-linked?

A: Journalists worldwide use common search engines to gather information about citizens and historical events. The news-gathering process may contain "holes" if information is de-linked from the search engine, allowing the reporters to discern inaccurate or incomplete interpretations of events. Despite the de-listing from the search engine, journalists may still access relevant data from original sources, such as courthouses or original archives from media outlets.

Q.: How does this European law affect Americans?

A: Some argue that the European law does not directly influence Americans. However, pressures to honor the right to be forgotten in the United States have sparked discussion about First Amendment protections and the enforcement of legal decisions made outside the United States. And some privacy experts have called for adoption of the right to be forgotten in the United States. Further, staunch advocates of the free flow of information indicate that removal of information in Europe alters the total availability of information for citizens and demonstrates European hegemony.

Q: Do United States citizens have an alternative for getting data removed from search engines?

A: Citizens in the United States can take legal action for data removal if they are victims of revenge pornography or they may utilize existing legal channels for libelous material. United States residents do not have access to the request forms found in Europe for search engine de-linking.

Q: Are search engine results protectable under the First Amendment?

A: Legal scholars continue to debate the appropriateness of First Amendment protection for search engine results. Some believe that the broad category of algorithmically generated speech merits protection. It has also been claimed that search engines demonstrate an editorial control over search engine results, suggesting First Amendment protection.

Q: Is the Right to Be Forgotten just a privacy concern?

A: While some may assert that the right to be forgotten is a privacy concern and that individuals should have the ability to undergo "digital redemption," the right to be forgotten may further be perceived as an issue that could imperil the free flow of information in society. By complicating the notion of the right to be forgotten with a privacy claim, scholars may lose sight of the broader issue of information integrity and the deleterious results of censorship and erasure of collective memory.

Q: How does the Right to Be Forgotten change the free flow of information in society?

A: The right to be forgotten can create information or "memory holes" in the inventory of retrievable information from the World Wide Web. When a European citizen petitions Google or another search engine for data removal and prevails in the request, the data is subsequently de-linked, making the information more difficult to find and potentially erasing it from the practical retrievability of information.

Appendix B

A Brief Timeline of the European Right to Be Forgotten

(A Provision of the General Data Protection Regulation)

1995: The European Union passed Regulation 95/46 known as the Data Protection Regulation governing the use of personally identifiable data about European Union citizens.

1998: Article published in *La Vanguardia* about a Spanish citizen in connection with a real estate transaction in connection with a bankruptcy filing.

2010: The first claim to the right be to forgotten issued under the Data Protection Regulation disputing the appropriateness of the *La Vanguardia* article.

2012: European Commission discloses draft of European Data Protection Regulation, incorporating the right to be forgotten in Article 17.

2014: *Google Spain v. Gonzalez* decided in the European Court of Justice, assigning Google and other search engines as data controllers.

2016: The General Data Protection Regulation was adopted by the Council of the European Union and adopted by the European Parliament.

2016: Google Transparency Report reveals right to be forgotten requests and defines the administrative process.

2016: The regulation went into force on May 24, 2016.

2017: French Authorities (CNIL) rule that Google must extend the right to be forgotten to all domains. Google files an appeal in July 2017.

2018: Companies must enter compliance with the General Data Protection Regulation by May 25, 2018.

Appendix C

The Free Speech Manifesto for the Digital Age: Seven Tenets to Preserve Information Flow in Democracy

Figure C.1: The Free Speech Manifesto for the Digital Age: Seven Tenets to Preserve Information Flow in Democracy.
***Source*: Courtesy of the author.**

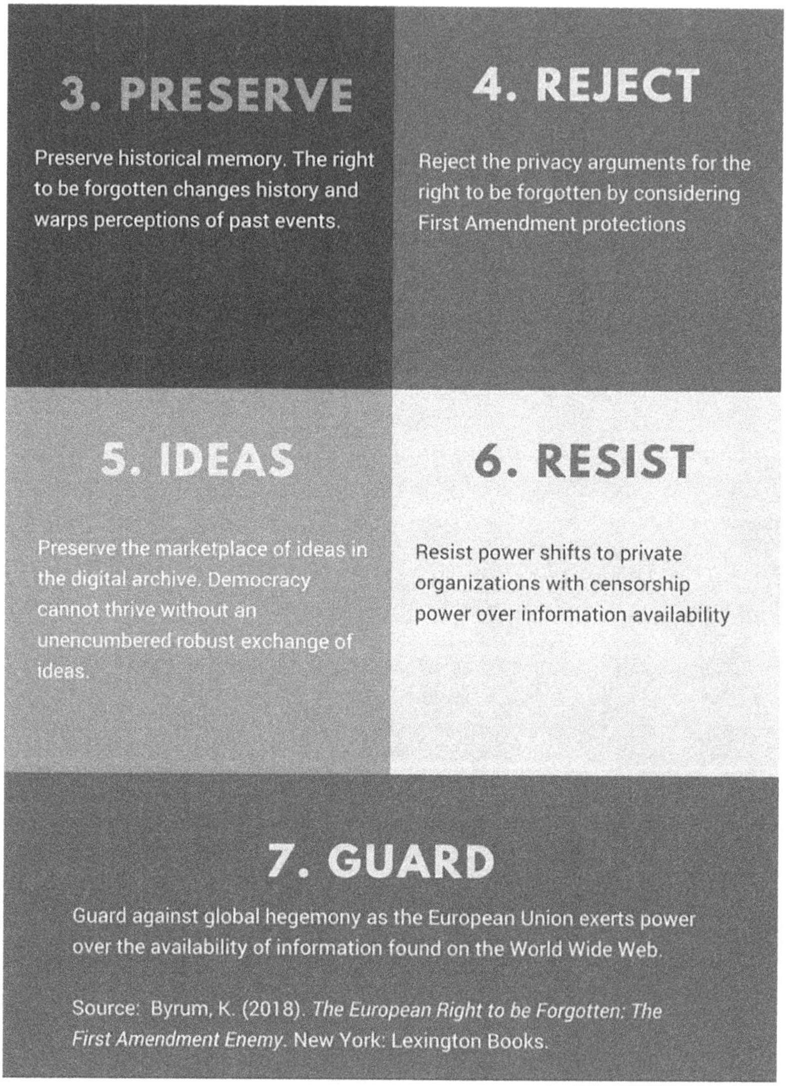

Figure C.1: The Free Speech Manifesto for the Digital Age: Seven Tenets to Preserve Information Flow in Democracy. (*continued*)

Appendix D

Media Consumption Test

Figure D.1: Media Consumption Test.
***Source*: Courtesy of the author.**

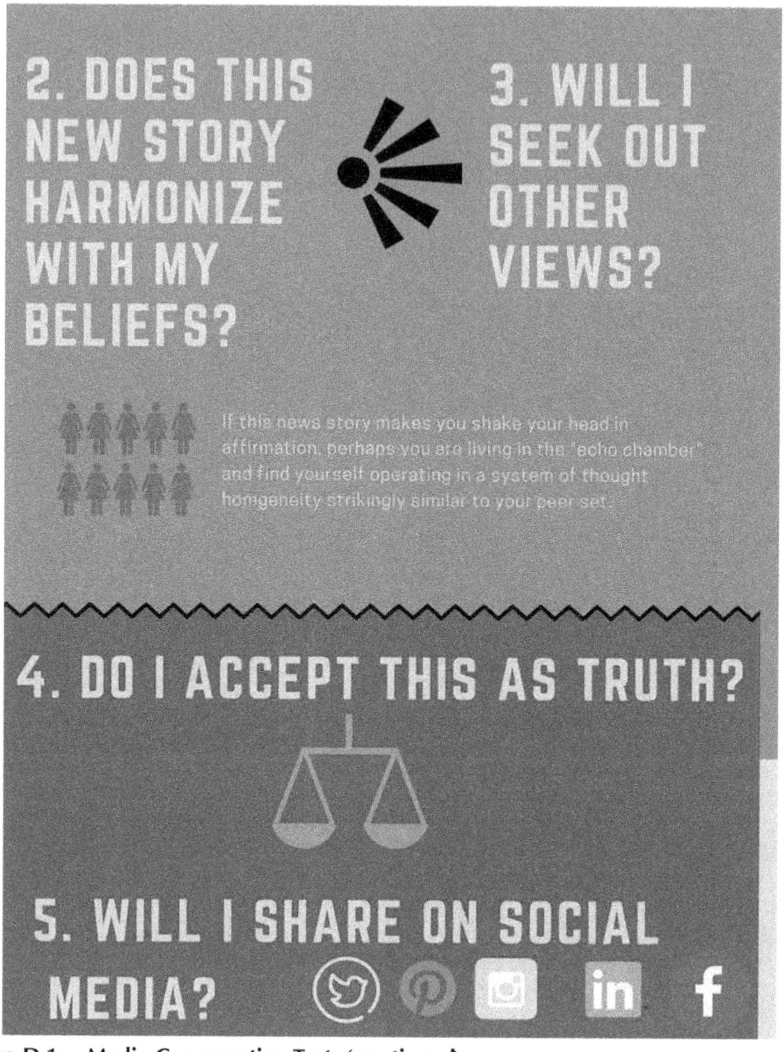

Figure D.1: Media Consumption Test. (*continued*)

References

Abrams v. United States, 250 U.S. 616 (1919). Retrieved from https://www.law.cornell.edu/supremecourt/text/250/616

Abrams, F. (2017). *The Soul of the First Amendment*. New Haven: Yale University Press.

Aiello, L., & Proffitt, J. (2008). VNR usage a matter of regulation or ethics? *Journal of Mass Media Ethics*, 23, 219–234.

Alphabet (2016). Form 10-K. Alphabet, Inc. Retrieved July 1, 2017, https://abc.xyz/investor/pdf/20151231_alphabet_10K.pdf

Amar, V., Ed. (2009). *The First Amendment Freedom of Speech. Its Constitutional History and the Contemporary Debate* (pp. 295). Amherst, NY: Prometheus Books.

Ambrose, M., & Ausloos, J. (2013). The right to be forgotten across the pond. *Journal of Information Policy*, 3, 1–23.

Ambrose, M., Friess, N., & Van Matre, J. (2012). Seeking digital redemption: The future of forgiveness in the Internet Age. *Santa Clara Computer and High Technology Law Journal*, 29.

Andrus, M. (2016a). The right to be forgotten in America: Have search engines inadvertently become consumer reporting agencies? *Business Law Today*, May, 1–4.

Andrus, M. (2016b). Constitutional issues in granting Americans a "Right to Dispute" personal information with search engines akin to the existing remedy afforded to Europeans via Europe's right to be forgotten. *Business Law Today*, November.

Andryushchenko, E. (2016). Right to be forgotten on the Internet in Europe and Russia. *Conhecimento & Diversidade, Niteroi*, (8), 15, 14–25.

Antani, R. (2015). The resistance of memory: Could the European's right to be forgotten exist in the United States? *Berkeley Technology Law Journal*, 30, 1173–1209.

Apuzzo, M., & LaFraniere, S. (2018). Thirteen Russians Indicted as Mueller Reveals Effort to Aid Trump Campaign. *New York Times*, February 16, 2018, retrieved https://nyti.ms/2C6XKm6.

Arendt, H. (1951). *The Origins of Totalitarianism*. New York: Harcourt.

Arendt, H. (1954). *Between Past and Future*. New York: Penguin.

Arthur, C. (2009). Wikipedia sued by German killers in privacy claim. *The Guardian* online. Retrieved from https://theguardian.com/technology/2009/nov.13/Wikipedia-sued-privacy-claim.

Balkin, J. (2012). The First Amendment is an information law policy. *Hofstra Law Review*, 41 (1).

Balkin, J. (2016). Cultural democracy and the First Amendment. *Northwestern University Law Review*, 110(5), 1053–1095.

Barron, J. (1967). Access to the press: A new first amendment right. *Harvard Law Review*, 80(8), 1641–1678.

Bass, A. (1969). Redefining the gatekeeper concept: A UN radio case study. *Journalism Quarterly*, 46, 69–72.

BBC (2017). List of BBC Webpages which have been removed from Google's search results. Retrieved from http://www.bbc.co.uk/blogs/internet/entries/131cf7ea-e3c1–4c72–8a10–43bd9b153753

Bernays, E. (1923*). Crystallizing Public Opinion*. New York: Ig Publishing.

Benjamin, S. (2013). Algorithms and speech. *University of Pennsylvania Law Review*, 161 (6), 1445–1494.

Bennett, S. (2012). The right to be forgotten: Reconciling EU and US perspectives. *Berkeley Journal of International Law*, 36(1), 161–195.

Bentham, J. (1843). The Works of Jeremy Bentham, vol. 4 (Panopticon, Constitution, Colonies, Codification). Retrieved http://oll.libertyfund.org/titles/bentham-the-works-of-jeremy-bentham-vol-4#lf0872–04_head_004

Bezanson, R. (1999). The developing law of editorial judgment, 78 *Nebraska Law Review*, 754.

Bhagwat, A. (2016). The democratic first amendment. *Northwestern University Law Review,* 110(5), 1097–1124.

Bing (2017). Request to block Bing search results in Europe, retrieved from https://www.bing.com/webmaster/tools/eu-privacy-request

Birmingham, P. (2007). A lying world order: Deception and the rhetoric of terror. *The Good Society*, 16(2), 32–37.

Birmingham, P. (2008). Elated citizenry: Deception and the democratic task of bearing witness. *Research in Phenomenology*, 38, 198–215.

Blackman, J. (2009). Omniveillance, Google, privacy in public, and the right to your digital identity: A Tort for recording and disseminating an individual's image over the Internet. *Santa Clara Law Review*, 49 (2), 313–392.

Blanchette, J., & Johnson, D. (2002). Data retention and the panoptic society: The social benefits of forgetfulness. *The Information Society*, 18, 33–45.

Blocher, J. (2008). Institutions in the marketplace of ideas. *Duke Law Journal*, 57(4).

Bode, L., & Jones, M. (2017). Ready to forget: American attitudes toward the right to be forgotten. *The Information Society*, 33:2, 76–85. doi:10.1080/01972243.2016.1271071

Bolton, R. (2015). The right to be forgotten: Forced amnesia in a technological age. *The John Marshall Journal of Information Technology & Privacy Law*, 31 (2), 133–144.

Bork, R. (1971). Neutral principles and some First Amendment problems. 47 *Ind. L.J.* 1, 20–21

Boshoff, A. (2012). Archive Fever: "order is no longer assured." *South African Journal of Philosophy*, 31(4).

Bracha, O. (2014). Folklore of informationalism: The case of search engine speech. *Fordham Law Review*, 82, 1629.

Bracha. O., & Pasquale, F. (2008). Federal Search Commission? Access, fairness, and accountability in the law of search. *Cornell Law Review*, 93, 1149–1209.

Brandenburg v. Ohio, 395 U.S. 444 (1969). Retrieved from https://supreme.justia.com/cases/federal/us/395/444/

Brazeal, G. (2011). Revisiting the marketplace of ideas. *Southern California Interdisciplinary Law Journal*, 21(1), 1–46.

Brock, G. (2016). *The Right to be Forgotten: Privacy and Media in the Digital Age*. London: I.B. Tauris & Co, Ltd.

Bunn, A. (2015). The curious case of the right to be forgotten. *Computer Law and Security Review*, 31, 336–350.

Burkert, H. (2007). Freedom of Information and Electronic Government in *Governance and Information Technology—From Electronic Government to Information Government* ed. Viktor Mayer-Shönberger and David Lazar, Cambridge, MA: MIT Press, p. 125–141.

Byrum, K. (2017). The European right to be forgotten: A challenge to the United States Constitution's first amendment and to professional public relations ethics. *Public Relations Review*, 43, 102–111. Retrieved from http://dx.doi.org/10.1016/j.pubrev.2016.10.010

Carbone, C. (2015). To be or not to be forgotten: Balancing the right to know with the right to privacy in the Digital Age. *Virginia Journal of Social Policy and the Law*, 22(3), 526–561.

Casey, A. J., & Olivera, F. (2011). Reflections on organizational memory and forgetting. *Journal of Management Inquiry*, 20, 305–310.

Chandler, J. (2007). Reclaiming the first amendment: Constitutional theories of media reform, A right to reach an audience: An approach to intermediary bias on the Internet. *Hofstra Law Review*, 35, 1095–1137.

Cheng, A. (2016). Forget about the right to be forgotten: How about a right to be different? *Auckland University Law Review*, 22.

Christopher Langdon v. Google Inc., et al. (2007) WL 530156, Civ. Act. No. 06–319-JJF (D. Del. February 20, 2007). Retrieved from http://www.internetlibrary.com/cases/lib_case458.cfm

Constitution Society (2017, January 12). Retrieved from http://www.constitution.org/rf/sedition_1798.htm

Consumer Watchdog (2015). News Release: Google's Failure to Offer "Right to Be Forgotten" in United States is Unfair and Deceptive, Consumer Watchdog Tells Federal Trade Commission in Complaint. Retrieved September from http://www.consumerwatchdog.org/newsrelease/google%E2%80%99s-failure-offer-%E2%80%98right-be-forgotten%E2%80%99-united-states-unfair-and-deceptive-consumer-.

Cook, H. (2017). Flagging the middle ground of the right to be forgotten: Combatting old news with search engine flags. *Vanderbilt Journal of Entertainment and Technology Law*. 20 (1), 1–50.

Cooper, E. (2016). Following in the European Union's footsteps: Why the United States should adopt its own "Right to be Forgotten" law for crime victims. *The John Marshall Journal of Information Technology & Privacy Law*, 32 (3), 185–210.

Cox Broadcasting Corporation v. Cohn (2017). *Oyez*. Retrieved August 7, 2017, from https://www.oyez.org/cases/1974/73–938

Cunningham, M. (2016). Free expression, privacy, and diminishing sovereignty in the Information Age: The internationalization of censorship. *Arkansas Law Review*, 69, 71–115.

Curtis, S., & Philpson, A. (2014). Wikipedia founder: EU's Right to Be Forgotten is "deeply immoral." Retrieved from http://www.telegraph.co.uk/technology/wikipedia/11015901/EU-ruling-on-link-removal-says-Wikipedia-founder.html

Cyber civil rights initiative (2017). Thirty-eight states and the District of Columbia have revenge pornography laws. Retrieved from https://www.cybercivilrights.org/revenge-porn-laws/

De Baets, A. (2016). A historian's view on the right to be forgotten. *International Review of Law, Computers and Technology*, 30 (1–2), 57–66.

DeIuliis, D. (2015). Gatekeeping theory from social fields to social networks. *Communication Research Trends*, 34(1), 4–23.

deMars, S., & Callaghan, P. (2016). Privacy and search engines: Forgetting or Contextualizing? *Journal of Law and Society*, 43(2), 257–284.

Derrida, J. (1995). *Archive Fever: A Freudian Impression*. Baltimore, MD: Johns Hopkins University Press.

Donovan, D. (2017). Mission scrubbed: Removal of Google search results has high bar. *South Carolina Lawyer's Weekly*, February 1.

Dulong de Rosnay, M., & Guadamuz, A. (2016). Memory hole or right to delist? *RESET* Online, doi:10.4000/reset.807

Edelman (2017a). Executive Summary: Edelman Trust Barometer 2017 Annual Global Study. Retrieved from http://www.edelman.com/executive-summary/

Edelman (2017b). News Release: 2017 Edelman Trust Barometer reveals global implosion of trust. Retrieved from http://www.edelman.com/news/2017-edelman-trust-barometer-reveals-global-implosion/

Edelman (2018). News Release: 2018 Edelman Trust Barometer reveals record-breaking drop in trust in the United States. Retrieved from https://www.edelman.com/news-awards/2018-edelman-trust-barometer-reveals-record-breaking-drop-trust-in-the-us

Electronic Frontier Foundation (2017). CDA 230: The most important law protecting Internet speech. Retrieved from https://www.eff.org/issues/cda230

European Commission news release (2015). Questions and answers—data protection reform. Retrieved from http://europa.eu/rapid/press-release_MEMO-15–6385_en.htm

European Convention on Human Rights (2017). Retrieved from http://www.echr.coe.int/Documents/Convention_ENG.pdf

European Union (2017). Charter of Fundamental Rights. Retrieved from http://eur-lex.europa.eu/legal-content/EN/TXT/?uri=CELEX:12012P/TXT

Ewan, S. in Bernays, E. (1923). *Crystallizing Public Opinion*. New York: Ig Publishing.

Facebook (2017). Terms of Service. Retrieved from https://www.facebook.com/terms.php

Feenstra, R. (2014). Communication ethics: Monitoring as a complement to self-regulation in the pursuit of transparency in the media industry. *Ramon Llull Journal of Applied Ethics*, 5, 113–124.

Fine, G.A., & Beim, A. (2007). Introduction: Interactionist approaches to collective memory. *Symbolic Interaction*, 30, 1–5.

Finigan, T. (2011). Into the memory hole: Totalitarianism and Mal d'Archive in Nineteen Eighty-Four and the Handmaid's Tale. *Science Fiction Studies*, 38, 435–459.

First Amendment Center (2017). Retrieved from http://www.firstamendmentcenter.org/madison/wp-content/uploads/2014/06/State-of-the-First-Amendment-2014-report-06–24–14.pdf

Fleischer, P. (2011). *Foggy thinking about the right to be forgotten*. Retrieved from http://peterfleisher.blogspot.com/2011/03/foggy-thinking-about-right-to-oblivion.html

Florida Star v. B. J. F. (1989). *Oyez*. Retrieved from https://www.oyez.org/cases/1988/87–329

Forde, A. (2015). Implications of the right to be forgotten. *Tulane Journal of Technology and Intellectual Property*, Fall, Vol. 18.

Foucault, M. (1991). *Discipline and Punish: The Birth of the Prison*. New York: Random House.

Founders Constitution (2017, January 12). Retrieved from http://press-pubs.uchicago.edu/founders/documents/amendI_speechs24.html

Friedrich, C., & Brzezinksi, Z. (1962). *Totalitarian Dictatorship and Autocracy*. New York: Frederick A. Praeger, Inc.

Gabrielson, T. (2009). James Madison's psychology of public opinion. *Political Research Quarterly*, 62 (3).

Ghosh, S. (2012). Informing and reforming the marketplace of ideas: The public-private model for data production and the first amendment. *Utah Law Review*, Vol. 2, 653–706.

Gibbs, S. (2016). Google to extend right to be forgotten to all its domains accessed in EU. *The Guardian*. Retrieved March 26, 2016, from http://www.theguardian.com/technology/2016/feb/11/google-extend-right-to-be-forgotten-googlecom.

Gilbert, F. (2015). The right to erasure or right to be forgotten: What the recent laws, cases, and guidelines mean for global companies. *Journal of Internet Law*, 18(8), 13–20.

Ginsberg, J. (2016). Global View: Europe's right to be forgotten pushed to new extremes after Belgian Court rules that individuals can force newspapers to edit archive articles. *Index on Censorship*, 45(3), 60–61. doi: 10.1177/0306422016670348

Gitlow v. New York, 268 U.S. 652 (1925).

Glazer, S. (2015). Right to be forgotten. *CQ Researcher*, 25(43), 1021–1022.

Global Internet User Survey, Internet Society (2012). The Global Internet User Summary Report, retrieved http://wayback.archive-it.org/9367/20170911022514/https://www.internetsociety.org/internet/global-internet-user-survey-2012

Google (2017). Frequently asked questions. Retrieved from https://support.google.com/transparencyreport/answer/7380434/

Google (2017). Search removal request under data protection law in Europe. Retrieved from https://transparencyreport.google.com/eu-privacy/overview

Google (2017). The Advisory Council to Google on the Right to be Forgotten. Retrieved from https://drive.google.com/file/d/0B1UgZshetMd4cEI3SjlvV0hNbDA/view

Google (2017). Transparency Report: European privacy requests for search removals. Retrieved from https://transparencyreport.google.com/eu-privacy/overview

Google (2017). Welcome to the Google Privacy Policy. Retrieved from https://www.google.com/policies/privacy/

Google (2018). Transparency Report: Search removals under European privacy law. Retrieved from https://transparencyreport.google.com/eu-privacy/overview.

Gordon, J. (1997). Mill and the marketplace of ideas. *Social Theory and Practice*, 23(2), 235–249.

Gorzeman, L., & Korenjof, P. (2017). Escaping the Panopticon over time: Balancing the right to be forgotten and freedom of expression in a technological architecture. *Philosophy of Technology*, 30, 73–92. doi: 10.1007/s13347–016–0238-y

Government in Sunshine Act, 5 USC 552b.

Grimmelmann, J. (2007). The structure of search engine law, *Iowa Law Review*, 93(1), 1–63.

Grimmelmann, J. (2008). The Google dilemma, *New York Law School Review*, 53, 939–950.

Grimmelmann, J. (2014). Speech engines. *Minnesota Law Review*, 98, 868–952.

Grunig, J., & Grunig, L. (1992). Models of Public Relations and Communications. In J. Grunig (Ed.), *Excellence in Public Relations Management and Communications and Communications Management* (pp. 285–324). Hillsdale, NJ: Lawrence Erlbaum Associates, Inc.

Gstrein, O. (2015). The cascade of decaying information: Putting the "Right to be Forgotten" in Perspective. *Computer and Telecommunications Law Review*, 21 (2), 40–48.

Guth, D., & Marsh, C. (2012). *Public Relations: A Values-Driven Approach.* Boston, MA: Allyn & Bacon, an imprint of Pearson.

Habermas, J. (1991). *The Structural Transformation of the Public Sphere*. Cambridge, MA: MIT Press.

Hanitzsch, T. (2011). Populist disseminators, detached watchdogs, critical change agents and opportunist facilitators: Professional milieus, the journalistic field and autonomy in 18 countries. *International Communication Gazette*, 73(6).

Healy, T. (2013). *The Great Dissent. How Oliver Wendell Holmes changed his mind and changed the history of free speech in America.* NY: Picador Paperback.

Hill, K. (2011). How the past haunts us in the Digital Age. *Forbes,* Oct. 4, 2011.

Hindman, M. (2009). *The Myth of Digital Democracy*. New Jersey: Princeton University Press.

Hoskins, A. (2011). Anachronisms of media, anachronisms of memory: From collective memory to a new memory ecology in Neiger, N., Meyers, O. & Zandenburg, E. (Eds.) *On Media Memory*. U.K.: Macmillan UK.

Hoskins, A. (2014). The right to be forgotten in post-scarcity culture in Ghezzi, A., Pereira, A., & Cesnick-Alujevic, L. (Eds). The ethics of memory in a Digital Age. U.K.: Palgrave Macmillan.House of Lords (UK). (2014). European Union Commission, EU Data Protection Law: "A Right to Be Forgotten," HL 40. PP 54–56.

Hurley v. Irish-American Gay, Lesbian and Bisexual Group of Boston, Inc. (1995). *Oyez*. Retrieved from https://www.oyez.org/cases/1994/94–749

Hurst, A. (2015). Data privacy and intermediary liability: Striking a balance between privacy, reputation, innovation and freedom of expression. *Journal of Internet Law*, 19(3), 9–21.

Instagram (2017). Terms of use. Retrieved https://help.instagram.com/47874555 8852511

International Association of Privacy Professionals (IAPP), A Brief History of Safe Harbor. Retrieved from The International Association of Privacy Professionals (IAPP)

Internet Archive (2017). Wayback machine. Retrieved from https://archive.org/web/

Internet Society (2012). Global Internet User Summary report. Retrieved from http://www.internetsociety.org/internet/global-internet-user-survey-2012.

Ivanovic, M. (2016). Lives rendered invisible: Bearing witness to human suffering. Nordic Journal of Applied Ethics, 59-74. doi: http://dx.doi.org/10.5324/eip.v10i2.1919

Janes, J. (2016). Forget me not: Is the appeal of the right to be forgotten more about anonymity or control? *American Libraries*. 47 (6), 30–31.

Jefferson, T. (2014). *Thomas Jefferson's Inaugural Addresses and First State of the Union Address*. London, England: Book Depository.

Jenkins, H. (2006). *Convergence Culture: Where Old and New Media Collide*. New York, NY: New York University Press.

Jones, M. (2016). *Control + Z: The Right to be Forgotten*. New York, NY: New York University Press.

Joo, T. (2014). The worst test of truth: The "marketplace of ideas" as faulty metaphor. *Tulane Law Review*, (89), 383–433.

Kamatali, J. (2016). The limits of the first amendment: Protecting American citizens' free speech in the era of the internet and the global marketplace of ideas. *Wisconsin International Law Review*, 33(4), 587–638.

Kaminksy, M., Rueben, M., Smart, W., & Grimm, C. (2017). Averting robot eyes. *Maryland Law Review*, 76, 983–1024.

Keane, J. (1991). *The media and democracy*. Cambridge: Polity Press.

Kelly, M., & Satola, M. (2017). The right to be forgotten. *University of Illinois Law Review*, 1.

Koien, G., & Oleshchuk, V. (2013). *Aspects of Personal Privacy in Communications: Problems, Technology and Solutions*. Aalborg, Denmark: River Publishers.

Koops, B. (2011). Forgetting footprints, shunning shadows. A critical analysis of the right to be forgotten in Big Data practice. *Scripted*, 8(3).

Kostov, N., & Schechner, S. (2017). EU Court to Rule on "Right to be Forgotten" Outside Europe; Case could determine whether France can force Google to apply the right to be forgotten across the globe. *Wall Street Journal Online*, July 19, 2017 retrieved from https://www.wsj.com/articles/eu-court-to-rule-on-right-to-be-forgotten-outside-europe-1500470225

Kulk, S., & Borgesius, F. (2014). Google Spain v. Gonzalez: Did the court forget about freedom of expression? *The European Journal of Risk Regulation*, (3), 389–398.

Lamparello, A. (2015). The internet is the new marketplace of ideas: Why Riley v. California supports net neutrality. *DePaul Journal Art, Technology & Intellectual Property Law*, XXV, 267–291.

Larson, R. (2013). Forgetting the First Amendment: How obscurity-based privacy and a right to be forgotten are incompatible with free speech. *Communication Law and Policy*. 18(91).

Lauchlan, S. (2014). Why Sir Tim Berners-Lee is frightened of the "right to be forgotten." Retrieved from http://dignomica.com/2014/10/08/sir-tim-berners-lee-frightened-right-forgotten/

Lechte, J., & Newman, S. (2012). Agamben, Arendt and human rights: Bearing witness to the human. *European Journal of Social Theory*, 15(4), 522–536.

Lee, E. (2016a). Recognizing rights in real time: The role of Google in the EU right to be forgotten. *University of California-Davis Law Review*, 48, 1017–1095.

Lee, E. (2016b). The right to be forgotten v. free speech. *I/S: A Journal of Law and Policy for the Information Society*, 12(85).

Legal Information Institute (2018). First Amendment. Retrieved from https://www.law.cornell.edu/constitution/first_amendment

Levinas, E. (1974). *Otherwise Than Being or Beyond Essence*. Trans. Lingis, A. Pittsburgh: Duquesne University Press.

Lewis, A. (1991). *Make No Law: The Sullivan Case and the First Amendment*. New York: Vintage Books.

Lewis, A. (2007). *Freedom for the thought we hate: A biography of the First Amendment*. New York: Basic Books, Perseus Books Group.

Lindsay, D. (2014). The "Right to be Forgotten" by search engines under data privacy law: A legal analysis of the Costeja ruling. *Journal of Media Law*, 6(2), 159–179.

Lippman, W. (1922). *Public Opinion*. Printed 2012. Scott's Valley, CA: Renaissance Classics.

Locascio, S. (2015). Forcing Europe to wear the rose-colored Google glass: The right to be forgotten and the struggle to manage compliance post Google Spain. *Columbia Journal of Transnational Law*, 54(1), 296–331.

Locke, J. (2001). *An essay concerning human understanding. [electronic resource]* Blacksburg, VA: Virginia Tech, 2001.

Lomas, N. (2014). Wikimedia attacks Europe's right to be forgotten ruling as threat to its mission. Retrieved from https://techcrunch.com/2014/08/06/wikimedia-rtbf/

Malgieri, G. (2016). Ownership of customer (Big) data in the European Union: Quasi-property as comparative solution. *Journal of Internet Law*, 3–17.

Manjoo, F. (2015). Right to be Forgotten online could spread. *New York Times*, August 5. Retrieved from https://www.nytimes.com/2015/08/06/technology/personaltech/right-to-be-forgotten-online-is-poised-to-spread.html?mcubz=3

Martin v. City of Stuthers, 319 U.S. 141 (1943).

Massaro, T., Norton, H., & Kaminski, M. (2017). SIRI-OUSLY 2.0: What artificial intelligence reveals about the First Amendment. *Minnesota Law Review*, 101, 2481–2525.

Mayer-Schönberger, V. (2009). *Delete: The virtue of forgetting in the Digital Age*. Princeton, N.J.: Princeton University Press.

McNealy, J. (2012). The emerging conflict between newsworthiness and the right to be forgotten. *Northern Kentucky Law Review*, 39(119).

Medium (2017). An open Letter to Google from 80 Internet scholars: Release RTBF Compliance. Retrieved from https://medium.com/@ellgood/open-letter-to-google-from-80-internet-scholars-release-rtbf-compliance-data-cbfc6d59f1bd#.oji3nlqgn

Meiklejohn, A. (1961). The first amendment is an absolute. *The Supreme Court Review*, 1961, 245–266.

Melvin v. Reid, 112 Cal. App. 285 (Cal. Ct. App. 1931).

Messenger, A. (2012). What would a "right to be forgotten" mean for media in the United States? *Communication Lawyer*, 29(1).

Miami Herald Publishing Company v. Tornillo (1973). *Oyez*. Retrieved from https://www.oyez.org/cases/1973/73–797

Mill, J. S. (2002). On liberty (Dover thrift ed.). Mineola, NY: Dover. (Original work published 1859).

Milton, J. (2014). *Aeropagitica*. Create Space Independent Publishing. (Original work published 1644).

Minora, L. (2017). U.S. courts should not let Europe's "Right to be Forgotten" force the world to forget. *Temple Law Review*, 89, 609–642.

Mitnick, K. (2017). *The Art of Invisibility*. New York, NY: Little, Brown and Company.

Moore, S. (2017). The right to be forgotten is the right to have an imperfect past. Opinion. The *Guardian*. Retrieved from https://www/theguardian.com.commentisfree/2017/aug/07/right-to-be-forgotten-data-protection-bill-ownership-identity-facebook-google.

National Constitution Center (2017, January 12). Retrieved from http://blog.constitutioncenter.org/2013/05/everything-you-ever-wanted-to-know-about-the-bill-of-rights/.

New York State Assembly (2017). Bill A5323. Retrieved from http://nyassembly.gov/leg/?default_fld=&leg_video=&bn=A05323&term=&Summary=Y&Text=Y

Nissenbaum, H. (2004). Privacy as contextual integrity. *Washington Law Review*, 79, 119–157.

Nissenbaum, H. (2010). *Privacy in Context: Technology, Policy, and the Integrity of Social Life*. Stanford, California: Stanford Law Books.

Ocasio, W., Mauskapf, M., & Steele, C. (2016). History, society and institutions: The role of collective memory in the emergence and evolution of societal logistics. *Academy of Management Review*. 41(4), 676–699.

Oltmann, S. (2016). Intellectual freedom and freedom of speech: Three theoretical perspectives. *Library Quarterly: Information, Community, Policy*, 86(2), 153–171.

Orwell, G. (1949). *1984*. New York, NY: Signet Classics.

Owings, L. (2015). The right to be forgotten. *Akron Intellectual Property Journal*, 9, 45–82.

Pacelle, R. (2017). Preferred Position Doctrine. The First Amendment Encyclopedia. Retrieved from https://mtsu.edu/first-amendment/article

Pacific Gas & Electric Company v. Public Utilities Commission of California (1986). *Oyez*. Retrieved from https://www.oyez.org/cases/1985/84–1044

Palko v. Connecticut (1937). *Oyez*. Retrieved from https://www.oyez.org/cases/1900–1940/302us319

Patriot Act (2001). Retrieved from https://www.justice.gov/archive/ll/highlights.htm

Peaslee, K. (2016). Does the United States need a "right to be forgotten"? *The Computer and Internet Lawyer*, 33(5), 12–24.

Peguera, M. (2016). The shaky ground of the right to be delisted. *Vanderbilt Journal of Entertainment and Technology*, 18(3), 507–561.

Peltz-Steele, R. (2015). The pond betwixt: Differences in the US-EU Data Protection/Safe Harbor negotiation. *Journal of Internet Law*, 19(1), 14–30.

Periera, A., Ghezzi, A., & Vesnic-Alujevnic (2014). Eds. *The Ethics of Memory in the Digital Age*. United Kingdom: Palgrave Macmillan.

Pew Research Center (November 2015). Global support for principle of free expression, but opposition to some forms of speech. Retrieved from http://www.pewglobal.org/2015/11/18/global-support-for-principle-of-free-expression-but-opposition-to-some-forms-of-speech/

Pew Research Center (February 2017). How Americans encounter, recall, and act upon digital news. Retrieved from http://www.journalism.org/2017/02/09/how-americans-encounter-recall-and-act-upon-digital-news/

Pike, G. (2016). The Right to be Forgotten. *Information Today*, 13.

Poritz, J. (2007). Who searches the searchers? Community privacy in the age of monolithic search engines. *Information Society*, 23, 383–389.

Post, R. (2011). Participatory democracy and free speech, 97 *Virginia Law Review*, 477, 482.

Powles, J., & Chaparro, E. (2015). How Google determined our right to be forgotten. *Guardian UK*, February 18, 2015. Retrieved July 5, 2017, http://www.theguardian.com/technology/2015/feb/18/the-right-to-be-forgotten-google-search.

Prosser, W. (1960). Privacy. *California Law Review*, 48, 383–389.

PRSA (2017). PRSA code of ethics. Retrieved from http://apps.prsa.org/AboutPRSA/ethics/codeenglish/

Pullen, J. (2017). Al Franken to Facebook: How could you not connect the dots in Russian election interference? *Fortune* magazine. Retrieved from http://fortune.com/2017/10/31/franken-facebook-russia-investigation/

Rand, R. (2010). Thoughts on "archive fever." *Communication and Critical/Cultural Studies*, 7(2), 207–211.

Red Lion Broadcasting v. FCC, 395 U.S. 367 (1969).

Reding, V. (2010). Why the EU needs new personal data protection rules. Retrieved from http://europa.eu/rapid/press-release_SPEECH-10–700_en.htm

Reno v. ACLU (1997). *Oyez*. Retrieved from https://www.oyez.org/cases/1996/96–511

Rhizome (2017). Artbase is Rhizome's archive of digital art retrieved from http://rhizome.org/art/artbase/

Ricouer, P. (2004). *Memory, History, Forgetting*. Chicago: University of Illinois Press.

Rosen, J. (2012). The deciders: The future of privacy and free speech in the age of Facebook and Google. *Fordham Law Review*, 80(4).

Royston, S. (2016). The right to be forgotten: Comparing U.S. and European approaches. *St. Mary's Law Journal*, 48, 253–275.

Rustad, M., & Kulevska, S. (2015). Reconceptualizing the right to be forgotten to enable transatlantic data flow. *Harvard Journal of Law and Technology*, 28(2), 351–417.

Sandefur, B. (2015). The best practice of forgetting. *Emory International Law Review*, 30(1), 85–114.

Sartor, G. (2015). The right to be forgotten in the draft data protection. *International Data Privacy Law*, 5(1), 64–72.

Schacter, D. (2001). *The Seven Sins of Memory: How the Mind Forgets and Remembers*. Boston, MA: Mariner Books, Houghton Mifflin Harcourt.

Schenck v. United States, 249 U.S. 47 (1919). Retrieved from https://www.oyez.org/cases/1900–1940/249us47

Schudson, M. (1990). *Origins of the Ideal of Objectivity in the Professions: Studies in the History of American Journalism and American Law* 1830–1940. New York and London: Garland Publishing.

Schudson, M. (1998). *The Good Citizen: A History of American Civic Life*. Free Press.

Schwartz, J. (2009). Two German killers demanding anonymity sue Wikipedia's parent, *New York Times*, Nov. 12, 2009. Retrieved from http://nytimes.com/2009/11/13/us/13wiki.html?_r=0.

Scott, M. (2016). Researchers uncover a flaw in Europe's tough privacy rules, *New York Times*, June 6, 2016, B3.

Search King, Inc. v. Google Technology, Inc. Case No CIV-02–1457-M. (2003). W.D. Okla., Jan. 13, 2003, retrieved http://www.internetlibrary.com/cases/ lib_case337.cfm

Shahin, S. (2016). Right to be forgotten: How national identity, political orientation and capitalist ideology structured a transatlantic debate on information access and control. *Journalism and Mass Communication Quarterly*, 93(2), 360–382.

Shapiro, L. (2017). Anatomy of a Russian Facebook Ad. *Washington Post*, November 1. Retrieved https://www.washingtonpost.com/graphics/2017/business/russian-ads-facebook-anatomy/?utm_term=.137f0ee4cc7c

Shoemaker, P., & Vos, T. (2009). *Gatekeeping Theory*. New York, NY: Routledge.

Siegel, P. (2014). *Communication Law in America*. Lanham, MD: Rowman & Littlefield.

Simpson, J., & Bates, S. (2015). Should the United States adopt the right to be forgotten? *CQ Researcher*, 25(43), 1025.

Society of Professional Journalists (2017). *SPJ Ethics Code*. Retrieved from http://www.spj.org/ethicscode.asp

Stenning, A. (2016). Gone but not forgotten: Recognizing the right to be forgotten in the U.S. to lessen the impact of data breaches. *San Diego International Law Journal*, 18, 129–160.

Stephens, M., Yoo, J., Mourao, R., Vu, H., Baresch, B., & Johnson, T. (2014). How apt are people to use smartphones, search engines and social media for news? Examining information acquisition tools and their influence on political knowledge and voting. *Journal of Information Technology and Politics*, 11, 383–396.

Stephens-Davidowitz, S. (2017). *Everybody else: Big Data, new data and what the Internet can tell us about who we really are*. New York, NY: Harper Collins.

Stewart, K. (2017). Looking backward, moving forward: What must be remembered when resolving the right to be forgotten. *Brooklyn Journal of International Law*, 42(2), 843–886.

Stute, D. (2015). Privacy almighty? The CJEU's judgment in Google Spain SL v AEPD. *Michigan Journal of International Law*, 36, 649–680.

Sunstein, C. (1993). *Democracy and the Problem of Free Speech*. New York, NY: The Free Press, a Division of Macmillan, Inc.

Svantesson, D. (2015). Limitless, borderless, forgetfulness? Limiting the geographical reach of the "Right to be Forgotten." *Oslo Law Review*, 2, 116–138. Retrieved from http://dx.doi.org/10.5617/oslaw2567

Tamo, A., & George, D. (2014). Oblivion, erasure and forgetting in the Digital Age. *Journal of Intellectual Property, Information Technology, and e-Commerce Law*. 5, 71–87.

Technopedia (2017). Citizen journalism. Retrieved from https://www.techopedia.com/definition/2386/citizen-journalism.

Telegraph (2017a). *Telegraph Stories affected by EU Right to be Forgotten*. Retrieved from http://www.telegraph.co.uk/technology/google/11036257/Telegraph-stories-affected-by-EU-right-to-be-forgotten.html

Telegraph (2017b). *Wikipedia founder: EU's Right to be Forgotten is "deeply immoral."* Retrieved from http://www.telegraph.co.uk/technology/wikipedia/11015901/EU-ruling-on-link-removal-deeply-immoral-says-Wikipedia-founder.html

Thomas v. Collins, 323 U.S. 516 (1945).

Thompson, B. (2005). The new visibility. *Theory, Culture and Society*, 22(6).

Timcke, S. (2017). The materials of memory: Tracing archives in communication studies. *Interactions: Studies in Communication and Culture*, 8(1), 9–20.

Tirosh, N. (2017). Reconsidering the "right to be forgotten": Memory rights and the right to memory in the new media era. *Media, Culture & Society*. 39(5), 644–660.

Toobin, J. (2014). The solace of oblivion: In Europe the right to be forgotten trumps the Internet. *The New Yorker*, September 29.

Townend, J. (2017). Data protection and the "right to be forgotten" in practice: A UK perspective. *International Journal of Legal Information*, 45(1).
Turner Broadcasting System, Inc. v. Federal Communications Commission (1997). *Oyez*. Retrieved from https://www.oyez.org/cases/1996/95–992
United Nations (2016). The Universal Declaration of Human Rights. Retrieved http://www.ohchr.org/EN/UDHR/Documents/UDHR_Translations/eng.pdf
United States v. Carolene Products Company (1938). *Oyez*. Retrieved from https://www.oyez.org/cases/1900–1940/304us144
U.S. v. Rumely, 345 U.S. 41 (Certiorari to U.S. Appeals for D.C. Circuit, 1953). Retrieved from https://supreme.justia.com/cases/federal/us/345/41/case.htmlU.S. Code, 47 U.S Code, 230. Retrieved from www.law.cornell.edu/uscode/text/47/230
Van der Sloot, B. (2017). Legal fundamentalism: Is data protection really a fundamental right? In Leenes, R., vanBrakel, R., & Gutwirth, S. Eds. (2017). *Issues in Privacy and Data Protection: Data Protection and Privacy: Invisibilities and Infrastructures*, 3–30.
Verheij, A. (2016). The right to be forgotten—A Dutch perspective. *International Review of Law, Computers & Technology*, 30 (1–2), 32–41.
Vogt, N., & Mitchell, A. (2016). Crowdfunded journalism: A small but growing addition to publicly driven journalism. Retrieved from http://www.journalism.org/2016/01/20/crowdfunded-journalism/
Volokh, E., & Falk, D. (2012). First Amendment protection for search engine search results—White Paper commissioned by Google (April 20, 2012). UCLA School of Law Research Paper No. 12–22. Retrieved from https://ssrn.com/abstract=2055364 or http://dx.doi.org/10.2139/ssrn.2055364
Walker, R. (2012). The right to be forgotten, 64 *Hastings Law Journal*, 257–271.
Warren, S., & Brandeis, L. (1890). The right to privacy. *Harvard Law Review*, 193.
Wechsler, S. (2015). The right to remember: The European convention on human rights and the right to be forgotten. *Columbia Journal of Law and Social Problems*, 49 (1), 135–165.
Weinstein, J. (2011). Participatory democracy as a central value of American free speech doctrine, 97 *Virginia Law Review*. 491, 493–497.
Werfel, E. (2016). What *organizations* must know about the "right to be forgotten." *Information Management*, March/April.
Werro, F. (2009). The right to inform v. The right to be forgotten: A Transatlantic clash. Georgetown University Center for Transnational Legal Studies Colloquium. Paper retrieved from http://ssrn.com/abstract=1401357
White, D. M. (1950). The gate keeper: A case study in the selection of news. *Journalism Quarterly*, 27, 383–390.
Whitman, J. (2004). The two Western cultures of privacy: Dignity versus liberty. *Yale Law Journal*, 111, 1151–1156.
Wikimedia (2017). Wikimedia Annual Report. Retrieved from https://wikimedia foundation.org/wiki/Annual_Report
Wikipedia (2017). Wolfgang Werle and Manfred Lauber. Retrieved from https://en.wikipedia.org/wiki/Wolfgang_Werl%C3%A9_Manfred_Lauber
Wu, T. (2012). Free speech for computers? *New York Times*, June 20, 2012, A29.

Wu, T. (2013). Machine Speech. *University of Pennsylvania Law Review*, 161, 1495–1533.

Yang, S., Quan-Haase, A., & Rannenberg, K. (2017) The changing public sphere on Twitter: Network structure, elites and topics of the #righttobeforgotten. *New Media and Society*, 19(2), 1983–2002.

YouTube (2017). Terms of Service. Retrieved from https://www.youtube.com/static?template=terms

Youm, K., & Park, A. (2016). The right to be forgotten in European law: Data protection balanced with free speech. *Journalism and Mass Communication Quarterly*, 93(2), 273–295.

Zittrain, J. (2010). *Reputation Bankruptcy*, in the Future of the Internet and How to Stop It. Retrieved from http://futureoftheinternet.org/2010/09/07/reputation-bankruptcy.

Index

About the Author

Kristie Byrum, PhD, APR, Fellow PRSA, is a global communications scholar with decades of rich experience as a professor, business executive, and journalist working in various roles in marketing, consulting, and entrepreneurism. She is an award-winning professional and honored member of the Public Relations Society of America's College of Fellows who actively works to advance strategic communication and its curricula on a global basis. A mass communications assistant professor at Bloomsburg University of Pennsylvania, Dr. Byrum teaches courses in media law, public relations, and mass communications research. She has presented her scholarly work in the United States, Switzerland, China, Norway, Canada, Greece, and Great Britain. She holds a PhD in rhetoric, communication, and information design from Clemson University, an MA in journalism and mass communications from the University of South Carolina, and a BA in journalism from The Pennsylvania State University.

CPSIA information can be obtained
at www.ICGtesting.com
Printed in the USA
BVHW07*0949190718
521978BV00001B/3/P

9 781498 549608